INTERNATIONAL CENTRE FOR MECHANICAL SCIENCES

COURSES AND LECTURES - No. 36

GORIMIR G. CHERNY

UNIVERSITY OF MOSCOW

LECTURES ON THE THEORY OF EXOTHERMIC FLOWS BEHIND SHOCK WAVES

COURSE HELD AT THE DEPARTMENT
OF HYDRO-AND GAS-DYNAMICS
JULY 1970

UDINE 1973

SPRINGER-VERLAG WIEN GMBH

ISBN 978-3-211-81168-9   ISBN 978-3-7091-2996-8 (eBook)

DOI 10.1007/978-3-7091-2996-8

PART 1

ONE – DIMENSIONAL NON – STEADY FLOWS WITH

INFINITELY THIN DETONATION WAVES

PART 2

STEADY GAS FLOWS AROUND BODIES WITH HEAT

RELEASE BEHIND THE BOW SHOCK WAVE

(Numerical methods)

PART 3

ONE–DIMENSIONAL NON–STEADY FLOWS WITH EXTENDED

HEAT RELEASE ZONE BEHIND SHOCK WAVE

P R E F A C E

*This book is the outgrowth of notes prepar
ed for a brief graduate course that I lectured in
Summer 1970 at the International Centre for Mechan-
ical Sciences in Udine (Italy). The lectures do not
pretend to provide a systematic presentation of all
the important problems of detonation theory. The
main content of the text is the theory of one-dimen
sional non-steady gas flows with detonation waves.*

*The lectures begin with discussion of con-
servation laws on infinitely thin detonation and de
flagration fronts. Then the simplest examples of ex
act solutions to gasdynamical equations describing
unsteady flows with detonation waves are given and
the problem of interaction of overdriven detonation
wave with rarefaction wave weakening is considered.*

*Different one-dimensional models of deton-
ation waves with extended heat release zone behind
the adiabatic shock wave are developed and some pro
blems concerning the stability of those detonation
waves in number of cases are discussed.*

In addition the lectures give a revue of

up-to-date results concerning numerical investigations of steady two-dimensional supersonic flow of
combustible gas mixtures around bodies.

The general pre-requisites of the book are
such that it can be used as complementary material
to textbooks dealing with detonation theory. A work
ing knowledge of general concepts of gasdynamics is
assumed.

The lectures naturally draw heavily on my
own research and that of my colleagues at the Insti
tute of Mechanics of Moscow University. This explains
a certain predominance of the material of this group
in some parts of the lectures. The author apologizes
for the unsystematic way in which references are gi
ven in the text and for many possible omissions.

During the preparation of the lectures I
have benefited from contact with many colleagues.
Specifically I wish to express my appreciation to
S.M. Gilinski and VA.A. Levin.

In particular the lectures 3 and 4 of this
text are entirely based on the material prepared by
S.M. Gilinski.

Udine, July 1970   G.G. CHERNY

INTRODUCTION

Since the time of early experiments by Veille, Berthelot, Mallard and Le Chatelier it is known that the combustion of gas mixtures filling long tubes may proceed in two diffe rent regimes.

In the slow combustion regime the gas burns in a flame front which propagates through the unburned mixture with velocity which does not exceed few meters per second. The velocity of flame front is determined by the transport processes : heat conduction and diffusion.

Along with this regime a quite different regime of flame propagation is possible. The compression and the heating of the combustible gas mixture which leads to a rapid intensification of chemical reactions in it, that is to the inflamation of a gas mixture, is due in this case to the passing through the gas of a sufficiently strong shock wave.

In a number of cases the combustion is then localized within a narrow zone behind the shock wave ; thus the velocity of flame propagation is in this case the same as that of shock wave and may be of the order of several kilometers per second. This mechanism of flame propagation got the name of detonation.

The experiments on detonation in tubes discerned

the constancy of the velocity of its propagation. During a long
period it was assumed that the front of the shock wave igniting
the gas is plane, the zone of chemical reactions behind the wave
has one-dimensional structure and the formation consisting of
the shock wave and combustion zone adjacent to it, i.e. the detonation wave, does not change with time. This imagination of
the detonation wave structure has been applied without sufficient
justification to the general case of its propagation.

If the thickness of the chemical reactions zone
is vanishingly small as compared with linear scale of the flow
field, it is possible in theoretical treatment of gas flows to
consider the detonation wave as a discontinuity surface with
instantaneous heat release due to the burning of the gas. On this
discontinuity the laws of mass, momentum and energy conservation
must hold.

Evidently if one postulates the possibility of
permanent existence of a narrow region behind the shock wave
where the heat release takes place, and if this region with the
shock wave ahead of it is modelled by a discontinuity surface,
then the solution of a large variety of gasdynamical problems
will not require to know details of the inner structure of the
detonation wave.

On the other hand it is evident that all these
solutions may be regarded only as asymptotic for time and lenght
scales, large enough in comparison with scales which are charac-

teristic for chemical reactions behind the detonation wave.
Therefore the question arises about the possibility of getting
these asymptotic solutions from different physically realistic
initial states, when the finite chemical reactions rates are
taken into account.

The importance of this question is emphasized by
the discovery at early investigations of detonation in tubes of
regimes when the propagating detonation wave is substanially
non one-dimensional (so called "spinning detonation") ; recent
experiments showed the non one-dimensional non-steady structure
of detonation waves in other cases as well. Moreover, theoretical investigations discovered and analyzed many cases of nonsta-
ble behaviour of detonation wave, when it is considered as one-
dimensional steady formation of shock wave and following it combustion zone.

The present lectures give a short review of basic
results in the theory of one-dimensional non-steady gas flows
with infinitely thin detonation waves (Lectures 1 and 2) and
analyze the main effects connected with finite thickness of the
chemical reaction zone under assumption of one-dimensional-
ity of the flow (Lectures 5 and 6). In addition a review of data
obtained by numerical methods in the problem of steady two-dimen
sional supersonic flow around bodies when the gas can burn behind the bow shock wave (Lectures 3 and 4) is given.

All lectures give a consequential treatment of the

subject ; but the lectures 3 and 4 may be read   independently
from others.

1.1. THE HUGONIOT CURVE FOR SHOCKS WITH HEAT RELEASE

The laws of mass, momentum and energy conserva-
tion across a discontinuity surface with instantaneous heat release may be written in the following form [1]

(1.1)
$$\varrho_1 \upsilon_1 = \varrho \upsilon = m$$

(1.2)
$$\varrho_1 \upsilon_1^2 + p_1 = \varrho \upsilon^2 + p$$

(1.3)
$$\frac{\upsilon_1^2}{2} + h_1 + Q = \frac{\upsilon^2}{2} + h.$$

Here p denotes pressure, ϱ –density, υ –velocity of the gas normal to the shock, m –mass flow rate across the
discontinuity surface, h –enthalpy of the unite of mass, which
is a function of thermodynamic parameters, say, pressure and den
sity, Q –the chemical energy released as heat when the gas mixture burns on the shock. These relations are written in the coordinate system related to the shock position and they differ
from the usual shock wave conditions only by the presence of the
term Q in the energy conservation law.

Let us deduce some consequences from the conser-
vation laws (1.1)-(1.3). Using the specific volume $V = \frac{1}{\varrho}$, instead of density ϱ , we obtain from (1.1) and (1.2)

$$\upsilon_1 = m V_1 \, , \quad \upsilon = m V \qquad (1.4)$$

$$p_1 + m^2 V_1 = p + m^2 V \qquad m^2 = \frac{p - p_1}{V_1 - V} . \qquad (1.5)$$

The last condition relates the velocity of shock propagation with pressures and densities on both sides of the shock. With given initial state of gas and m this relation determines in the p, V plane a straight line known as Michelson-Rayleigh line.

The conservation laws (1.1)-(1.3) give also the following relation

$$h(p, V) - h_1(p_1 V_1) = \frac{p - p_1}{2} (V_1 + V) + Q . \qquad (1.6)$$

This relation is called the Hugoniot shock adiabate. With p_1, V_1 given the shock adiabate determines the admissible values of pressure and specific volume in the combustion products behind the shock.

The graph of the shock adiabate is shown on Fig. 1. It is easy to show that the Hugoniot curve for flows with heat addition lies above the usual shock adiabate.

Let us assume the gas on both sides of the shock to be perfect with the same adiabatic exponent γ . In this case

$$h(p, V) = \frac{\gamma}{\gamma - 1} \, p V .$$

Then (1.6) gives :

(1.7) $$\left(\frac{\gamma+1}{\gamma-1}\ V - V_1\right)p - \left(\frac{\gamma+1}{\gamma-1}\ V_1 - V\right)p_1 = 2\,Q\ .$$

For an usual shock with same p_1 , V_1 and V

(1.8) $$\left(\frac{\gamma+1}{\gamma-1}\ V - V_1\right)p^* - \left(\frac{\gamma+1}{\gamma-1}\ V_1 - V\right)p_1 = 0\ .$$

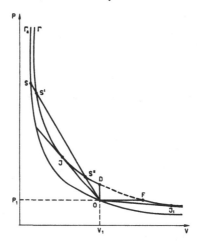

Fig. 1.

Subtracting (1.8) from (1.7) one gets

$$\left(\frac{\gamma+1}{\gamma-1}\ V - V_1\right)\left(p - p^*\right) = 2\,Q\ .$$

In the range $V > \dfrac{\gamma-1}{\gamma+1}\ V_1$ i.e. for all V obtainable when the gas passes the shock wave, the inequality $p > p^*$ holds. Thus, the Hugoniot curve with $Q > 0$ lies above the usual shock adiabate.

Depending on the position of the Michelson-Rayleigh straight line, which determines the velocity of shock propagation

$$\mathfrak{D} = \upsilon_1 = V_1 \sqrt{\frac{p-p_1}{V_1-V}} \qquad (1.9)$$

the Hugoniot curve is divided into several parts (Fig. 1).

On the $\mathfrak{D}F$ part the inequality $\frac{p-p_1}{V_1-V} < 0$ holds. This means the imaginarily value for the velocity of shock propagation ; therefore the part $\mathfrak{D}F$ of Hugoniot curve does not correspond to any physically possible state of gas.

The part of the curve down of the point F corresponds to normal slow combustion or deflagration regimes.

On the part upper to the point $\mathfrak{D}$ the combustion is followed by increase of pressure and density what corresponds to detonation or fast combustion regimes.

At the point p_1,V_1 the shock adiabate with $Q = 0$ and the Poisson's adiabatic curve are tangent ; consequently, the slope of the tangent to the shock adiabate at this point determines the speed of sound in the initial state. The slope of this tangent is less than the slope of any straight line connecting the initial state point with an arbitrary point on the curve Γ lying higher than point $\mathfrak{D}$. Therefore the corresponding shocks propagate through the gas with speeds exceeding the sonic speed in the initial gas mixture.

The part of the curve Γ up from the point $\mathcal{D}$ determines a continuous set of regimes of detonation propagating with velocities $\mathcal{D} = v_1$, depending on the slope of the Michelson straight line. Among these velocities there is the smallest velocity $\mathcal{D}_{min}$ which corresponds to the point $\mathcal{J}$ at which the Michelson line is tangent to the Hugoniot curve Γ.

Let us consider the variation of gas state in the layer adjacent to the shock wave, assuming the motion inside of this layer to be steady in the coordinate system related to the shock and one-dimensional. The front of a detonation wave is an usual shock wave propagating through undisturbed gas. In this shock the gas is compressed and heated. The state of the gas immediately behind the shock wave corresponds to the point S on the adiabatic curve Γ_a. In the compressed gas the chemical reaction starts in the course of which the state of the gas is described by the points of Michelson's line passing through point S. Let us assume that the heat is added to the gas continuously ; then the state of the gas in the combustion zone changes along the straight line SS', where the point S' lies on the curve Γ corresponding to the release of the whole heat Q. The lower point S'' at which the Michelson line crosses the curve Γ is unattainable due to the fact that to reach this point one has to move along the sector $S'S''$ which corresponds to heat release exceeding Q. The continuous movement from the initial state O to the state S'' can not be realized due to the lack of any transport mechanism which could produce so high velocities of the combustion front propagation through the gas. Therefore the detonation waves are described by the part of the Hugoniot curve lying above the point $\mathcal{J}$. This point is called the Chapman-Jouget point. At the CH-J. point the velocity of the gas v is equal to the local sound speed a. At points lying above the point $\mathcal{J}$ the gas velocity v is less than the speed of sound a, at points under $\mathcal{J}$ $v > a$.

Let us prove this statement. Along the Γ curve from (1.5) and (1.3) we get

$$dp + m^2 dV = (V_1 - V) dm^2 \qquad (1.10)$$

and

$$dh + m^2 V dV = \frac{V_1^2 - V^2}{2} dm^2. \qquad (1.11)$$

Making use of the relation

$$T ds = dh - V dp$$

we obtain that along the Hugoniot curve the following relation takes place

$$T ds = \frac{(V - V_1)^2}{2} dm^2. \qquad (1.12)$$

At the point $\mathfrak{z}$ the value of m^2 has a minimum , i.e.

$$\frac{dm^2}{dp} = 0 .$$

It follows from (1.12) that at this point $\left(\dfrac{\partial S}{\partial p}\right)_{\Gamma} = 0$ and entropy $\mathfrak{d}$ has a minimum.

If we write the differential dV in the form

$$dV = \left(\frac{\partial V}{\partial p}\right)_{S} dp + \left(\frac{\partial V}{\partial S}\right)_{p} d\mathfrak{d}$$

the relations (1.10) and (1.12) will give

$$(1.16) \qquad 1 + m^2 \left(\frac{\partial V}{\partial p}\right)_{S} = (V_1 - V)\left[1 - \frac{m^2(V_1 - V)}{2T}\left(\frac{\partial V}{\partial S}\right)_{p}\right]\frac{dm^2}{dp} .$$

It follows immediately that at points where $\dfrac{dm^2}{dp} = 0$

$$1 + m^2 \left(\frac{\partial V}{\partial p}\right)_{S} = 1 - \frac{v^2}{a^2} = 0$$

i.e. $v = a$. Moreover at these points

$$(1.17) \qquad\qquad \frac{d}{dp}\,\frac{v^2}{a^2} = -m^2\left(\frac{\partial^2 V}{\partial p^2}\right)_{S} .$$

Since $\left(\dfrac{\partial^2 V}{\partial p^2}\right)_{S} > 0$, it follows that $\left(\dfrac{d}{dp}\,\dfrac{v^2}{a^2}\right)_{\mathfrak{z}} < 0$, and this means that above the point $\mathfrak{z}$ $v < a$ and below it $v > a$.

The analogous consideration shows that $v < a$ on

the part $F \gamma_1$ and $\upsilon > a$ under the point γ_1.

Let us take note of a circumstance which is of importance for the following. In the case of a detonation wave corresponding to a point above γ $\upsilon_1 > a_1$, $\upsilon < a$. Therefore in the case of a detonation wave quite similar to the case of an ordinary shock wave only two kinds of outgoing small perturbations may exist, namely : acoustical perturbations propagating with velocity $\upsilon + a$ and entropy variations propagating with velocity υ -both in the flow direction (Fig. 2a). In the case of normal combustion front on the part $F \gamma_1$ $\upsilon_1 < a_1$, $\upsilon < a$. Consequently in addition to perturbations of the same nature as in the case of detonation wave in the case of the combustion front there is a third kind of outgoing perturbations : acoustical perturbations with velocity of propagation $a_1 - \upsilon_1$ in the direction opposite to the direction of the oncoming flow (Fig. 2b).

Since the number of shock conditions equals to three, it is evident that only these three conditions following from the conservation laws would be sufficient in the case of a detonation wave or a shock wave and would be not sufficient in the case of a normal condition front to determine the outgoing perturbations and the variation of the shock velocity when the perturbations coming to the shock are given. Thus the normal combustion front belongs to the so called non-evolutionary discontinuities. If a gasdynamical problem concerning flows with

combustion front has to be solved one additional condition on
the front or a condition replacing it must be given.

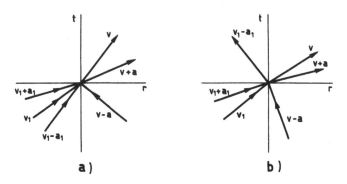

a)   b)

Figs. 2a and 2b

For a normal combustion front which propagates due to the molecular transport processes this condition is given by prescribing
a priori the velocity of front propagation. Theoretically this
velocity may be found through considerations of the internal
structure of the combustion zone.

1.2. EXAMPLES OF GAS FLOW WITH DETONATION WAVES

Let us consider some important examples of one-
dimensional gas flows with propagating detonation waves.

One-dimensional gas flows are described by the
following set of equations

$$(2.1a) \qquad \frac{\partial \varrho}{\partial t} + \frac{\partial \varrho v}{\partial r} + (\nu - 1)\frac{\varrho v}{r} = 0$$

$$\frac{\partial \upsilon}{\partial t} + \upsilon \frac{\partial \upsilon}{\partial r} + \frac{1}{\varrho} \frac{\partial p}{\partial r} = 0$$

$$\frac{\partial}{\partial t} \frac{p^{1/\gamma}}{\varrho} + \upsilon \frac{\partial}{\partial r} \frac{p^{1/\gamma}}{\varrho} = 0.$$

(2.1b)

Here t denotes time, r – is the linear coordinate, $\nu = 1, 2, 3$ respectively for flows with plane, cylindrical and spherical waves.

The parameters of the gas on both sides of the shock with heat addition are related by the conditions

$$\varrho_1 (\upsilon_1 - c) = \varrho (\upsilon - c)$$

$$\varrho_1 (\upsilon_1 - c)^2 + p_1 = \varrho (\upsilon - c)^2 + p$$

(2.2)

$$\frac{(\upsilon_1 - c)^2}{2} + \frac{\gamma}{\gamma - 1} \frac{p_1}{\varrho_1} + Q = \frac{(\upsilon - c)^2}{2} + \frac{\gamma}{\gamma - 1} \frac{p}{\varrho}.$$

Here c is the velocity with which the discontinuity propagates in the space.

Solving eqs. (2.2) we obtain

$$\upsilon - \upsilon_1 = \frac{1}{\gamma + 1} \left[\frac{a_1^2}{\upsilon_1 - c} - (\upsilon_1 - c) \pm \sqrt{\left[\frac{a_1^2}{\upsilon_1 - c} - (\upsilon_1 - c) \right]^2 - 2(\gamma^2 - 1) Q} \right]$$

$$p = p_1 + \varrho_1 (\upsilon_1 - c)/(\upsilon_1 - \upsilon)$$

(2.3)

$$\varrho = \frac{\varrho_1 (\upsilon_1 - c)}{\upsilon - c}.$$

If the shock propagates through the gas with supersonic velocity
the sign "+" in the expression for υ corresponds to detonation
waves ; if the velocity of shock propagation is subsonic the
sign "−" corresponds to the normal combustion front.

In the coordinate system related to the gas ahead
of the shock ($\upsilon_1 = 0$) eqs (3.2) may be given the following form

$$\upsilon = \frac{a_1}{\gamma+1} \; \frac{1-q \pm \sqrt{(1-qq_3)\left(1-\dfrac{q}{q_3}\right)}}{\sqrt{q}}$$

(2.4)
$$p = p_1 + \frac{\varrho_1 a_1 \upsilon}{\sqrt{q}}$$

$$\varrho = \frac{\varrho_1}{1 - \dfrac{\upsilon\sqrt{q}}{a_1}} .$$

Here $q = \dfrac{a_1^2}{c^2}$, q_3 denotes the value of q, which corresponds
to the velocity of the heat release front in the Chapman–Jouget
regime. This value is given by the equation

(2.5)
$$\frac{a_1}{\gamma+1} \frac{1-q_3}{\sqrt{q_3}} = \pm \left(2 \frac{\gamma-1}{\gamma+1} Q \right)^{1/2} .$$

The upper sign corresponds to the fronts propagating through the
gas with supersonic speed, the lower − with subsonic speed.

For the parameters of the gas behind the Ch.–J.

wave we have

$$v_3 = \frac{c_3}{\gamma+1}(1-q_3), \quad p_3 = \varrho_1 c_3^2 \frac{\gamma+q_3}{\gamma(\gamma+1)},$$

$$\varrho_3 = \varrho_1 \frac{\gamma+1}{\gamma+q_3}. \tag{2.6}$$

Along with the shock conditions solving a particular problem requires to take into account initial and boundary conditions which are different for different problems. If, for instance, the detonation wave propagates through uniform gas initially at rest from the point in the symmetry centre, the boundary conditions would be $v(0,t)=0$, $v(\infty,t)=v_1$, $p(\infty,t)=p_1$, $\varrho(\infty,t)=\varrho_1$. Initial conditions would be $v(r,0)=0$, $p(r,0)=p_1$, $\varrho(r,0)=\varrho_1$.

To treat the problems on perfect gas flows it is convenient to replace the thermodynamic variables p,ϱ in Eqs. (2.1) by a and s, where $a=(\gamma p/\varrho)^{1/2}$ is the speed of sound and $s=c_p \ln(p^{1/\gamma}/\varrho)$ is the gas entropy. By simple transformation we get

$$\frac{\partial}{\partial t}\left(v+\frac{2}{\gamma-1}a\right)+(v+a)\frac{\partial}{\partial r}\left(v+\frac{2}{\gamma-1}a\right)-\frac{a^2}{\gamma-1}\frac{\partial}{\partial r}\frac{s}{c_p}+(\nu-1)\frac{va}{r}=0$$

$$\frac{\partial}{\partial t}\left(v-\frac{2}{\gamma-1}a\right)+(v-a)\frac{\partial}{\partial r}\left(v-\frac{2}{\gamma-1}a\right)-\frac{a^2}{\gamma-1}\frac{\partial}{\partial r}\frac{s}{c_p}-(\nu-1)\frac{va}{r}=0$$

$$\frac{\partial s}{\partial t}+v\frac{\partial s}{\partial r}=0.$$

Let us introduce new independent variables $\lambda = \dfrac{r}{t}$ and $\tau = \ell n t$. The transformed equations have the form

$$\frac{\partial}{\partial \tau}\left(v + \frac{2a}{\gamma-1}\right) + (v+a-\lambda)\frac{\partial}{\partial\lambda}\left(v+\frac{2a}{\gamma-1}\right) - \frac{a^2}{\gamma-1}\frac{\partial}{\partial\lambda}\frac{s}{c_p} + \frac{(\nu-1)va}{\lambda} = 0$$

(2.7)
$$\frac{\partial}{\partial \tau}\left(v - \frac{2a}{\gamma-1}\right) + (v-a-\lambda)\frac{\partial}{\partial\lambda}\left(v-\frac{2a}{\gamma-1}\right) - \frac{a^2}{\gamma-1}\frac{\partial}{\partial\lambda}\frac{s}{c_p} - \frac{(\nu-1)va}{\lambda} = 0$$

$$\frac{\partial s}{\partial \tau} + (v-\lambda)\frac{\partial s}{\partial\lambda} = 0.$$

The system (2.7) has three families of characteristics. Along the characteristics of two first families (acoustical characteristics) the following conditions hold

$$\dot{\lambda} + \lambda - v = \pm a$$

(2.8)
$$\frac{1}{\gamma}\frac{\dot{p}}{p} \pm \frac{\dot{v}}{a} + \frac{(\nu-1)v}{\lambda} = 0$$

$$\left(\frac{1}{\gamma}\frac{\dot{p}}{p} \equiv \frac{2}{\gamma-1}\frac{\dot{a}}{a} - \frac{\dot{s}}{(\gamma-1)c_p}\right).$$

Along the characteristics of the third family (particles pathes) we have

(2.9)
$$\dot{\lambda} + \lambda - v = 0 \qquad \dot{s} = 0.$$

The point above denotes differentiation on τ along the characteristics.

We consider the following problem on the gas flow behind the heat release front. The uniform combustible gas initially at rest filling the whole space, is ignited at some instant at a point(along a straight line or a plane) in such manner that a detonation wave forms. Simultaneously with the ignition from the same point (straight line or plane) a spherical (cylindrical, plane) piston starts to expand with constant speed v_n The problem consists in the determination of the developing gas flow.

The formulation of the problem does not include any quantity which dimension is lenght or time : among the given parameters the dimensional quantities are only p_1, ϱ_1 (or a_1 and $\mathfrak{s}_1$), Q and v_n. As a consequence the gas flow must be self-similar, i.e. all parameters of the moving gas must depend only on one variable $\lambda = \dfrac{r}{t}$. The velocity of discontinuities separating one flow region from another must be constant.

Eqs. (2.7) under condition of self-similarity of gas motion become ordinary differential equations

$$(v + a - \lambda)\, \frac{d}{d\lambda}\left(v + \frac{2a}{\gamma-1}\right) + \frac{(\nu-1)va}{\lambda} = 0$$

$$(v - a - \lambda)\, \frac{d}{d\lambda}\left(v - \frac{2a}{\gamma-1}\right) - \frac{(\nu-1)va}{\lambda} = 0 \qquad (2.10)$$

$$\frac{d\mathfrak{s}}{d\lambda} = 0 \,.$$

(The possibility to have from the third equation $\upsilon = \lambda$ is not used because in this case the two first equations are contradictory).

The third equation has an integral δ = const, so the motions studied must consist of isentropic regions.

Let us consider first plane waves for which $\nu = 1$. In this case along with the obvious solution

$$\upsilon = const, \quad a = const, \quad \delta = const$$

which corresponds to the translational motion of the gas as a whole, there are the following solutions

$$\upsilon + a - \lambda = 0 \qquad \upsilon - \frac{2a}{\gamma - 1} = const \qquad \delta = const$$

and

$$\upsilon - a - \lambda = 0 \qquad \upsilon + \frac{2a}{\gamma - 1} = const \qquad \delta = const$$

which correspond to centered Riemann waves propagating through the gas in one or another side respectively.

Evidently in the self-similar motion with plane waves the heat release front which propagates with supersonic speed through the gas ahead of it cannot be precoursed by a shock wave or by a centred Riemann wave whose velocity relative to the gas behind them is less or equal to that of the sound. Thus the heat release front propagates through gas which is at rest. If this front is an overdriven detonation wave its speed relatively to

the gas behind it is subsonic and it is again obvious that the
space between the wave and the piston must be filled by uniform
flow : no shock wave or centred Riemann wave is possible in this
region. The velocity of the detonation wave is in this case determined by the piston velocity, which has to be in excess of
the gas velocity behind a Chapman-Jouget wave.

If the detonation wave propagates in the Ch.-J.
regime along with the uniform flow behind it another type of
flow between the wave and the piston is possible. Actually the
velocity of Ch.-J. wave relatively to the gas behind it coincides
with the sound speed. Due to this fact the rear side of the Ch.-
J. wave may be simultaneously the head side of a centred Riemann
wave of arbitrary strenght ; between the rear front of this
Riemann wave and the piston the region with uniform flow parameters establishes, where the gas has the same velocity as the pis
ton. The piston velocity in this case lies below the value of the
gas velocity behind Ch.-J. wave ; in particular the piston velocity may be equal to zero or even it may be negative.

Fig. 3 shows the distributions of gas parameters
for the flow with a Ch.-J. wave when the piston velocity equals
zero.

It follows from what was told previously that
with given constant piston velocity the solution of the piston
problem with detonation wave is determined uniquely.

Let us note that the rear front of the Ch.-J.

wave is a characteristic for the flow behind it : on this front
both characteristic relations (2.8) are satisfied.

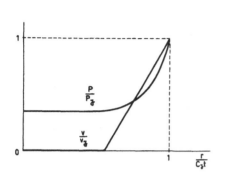

Therefore even in non self-
similar motion the Ch.-J.
wave progressing through uniform gas at rest may be followed by arbitrary (not neces
sarily centred) Riemann wave.
As a consequence, once being
established the Ch.-J. wave
will continue to exist even

Fig. 3

in the case when the piston which supports the propagation of
the wave will decelerate in arbitrary manner and finally will
stop completely. This is what is meant when one tells on "self-
supporting" detonation wave.

If we assume that a front of fast combustion propagates from the piston into the gas then the solution of the
problem with piston velocity given will not be unique. This is
due to the fact that the velocity of fast combustion front relative to the gas behind it exceeds the sound speed and, as a consequence, between the piston and the front a shock wave or a
centred rarefaction wave may be placed whose strength may be
chosen in some range arbitrarily.

Now we turn to the analogous problem for spheric-
al or cylindrical waves. In these cases Eqs. (2.10) do not pos-

sess simple integrals which would be sufficient to solve the problem and it is necessary to analyze the field of their integral curves. It is convenient to make use of new variables V and z taken as below [2]

$$v = \lambda V, \quad a^2 = \lambda^2 z.$$

The first two equations take then the form

$$\frac{dz}{dV} = \frac{z\left[2(V-1)^2 + (\nu-1)(\gamma-1)V(V-1) - 2z\right]}{V(V-1)^2 - \nu Vz} \quad (2.11a)$$

$$\frac{d\ln\lambda}{dV} = \frac{z - (V-1)^2}{V(V-1)^2 - \nu Vz}. \quad (2.11b)$$

Fig. 4a shows the integral curves of Eq $(2.11a)$ for $\nu = 3$ $\left(\gamma = \frac{5}{3}\right)$. The sense of growing λ is shown by arrows on the curves. On the parabola $z = (V-1)^2$ the value of λ has an extremum. Therefore it is impossible to cross this parabola along integral curves at any point excluding the singular point $A(0,1)$.

In the problems under consideration λ must vary from its value on the piston to infinity. Since on the piston $\frac{r}{t} = v_n$ the piston corresponds to the straight line $V = 1$ in the V, z plane. In that part of the space which is not crossed by the detonation wave yet the gas is at rest. This state corresponds to the points of V-axis from point $O(0,0)$ till point $H(0,q)$. Here as previously $q = \frac{a_1^2}{c^2} < 1$. From two first Eqs.$(2.2)$ with $v_1 = 0$ it follows that the point behind the shock must lie on the parabola

(2.12) $$z = (q + \gamma V)(1 - V).$$

With account of this the last Eq. (2.2) gives for the same point the following condition

(2.13) $$z = V(1-V) \; \frac{1 + \frac{\gamma-1}{2} V + \frac{\gamma(\gamma-1)Q}{a_1^2}}{V + \frac{(\gamma-1)Q}{a_1^2}}.$$

The points where the parabola (2.12) intersects the curve (2.13) determine in the V, z plane the state behind the shock. The range of all possible values of V_s, z_s is bounded besides of the part OA of the z -axis by the curve (2.13) with $\frac{Q}{a_1^2} = 0$ (what corresponds to conditions behind an adiabatic shock), i.e. by the parabola

$$z = (1-V)\left(1 + \frac{\gamma-1}{2} V\right)$$

and by the parabola

$$z = \gamma V(1-V)$$

corresponding to $\frac{Q}{a_1^2} = \infty$ in the Eq. (2.13) that is to an infinitely strong wave. Both these parabolas are shown by dotted lines on the Fig. 4a. Fig. 4b shows the mutual position of curves (2.12) and (2.13) in the region OAE. For any value of q and a sufficiently small value of Q the parabola (2.12) intersects the curve (2.13) at two points.

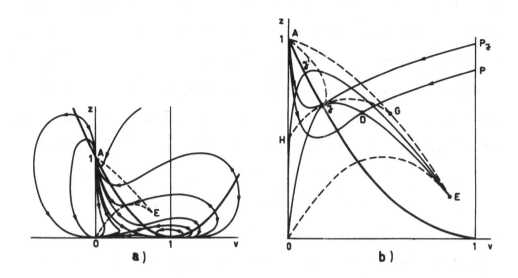

Fig. 4a and 4b

At some value of Q these two points merge into
one situated on the parabola

$$z = (V-1)^2 \qquad (2.14)$$

what evidently corresponds to the Ch.-J. regime, for in this
case $c - v = a$.

With given q and large Q the curves (2.12) and
(2.13) have no common points. If there are two points of intersection, that which lies above parabola (2.14) corresponds to
overdriven detonation, and that which lies under this parabola
corresponds to fast combustion front.

Let us assume that the quantity $\dfrac{Q}{a_1^2}$ is given,
that is the curve $O\mathfrak{J}\mathcal{E}$ on Fig. 4b is known. Through each point

$\mathfrak{D}$ of this curve on the $\mathfrak{J}\mathcal{E}$ section an integral curve passes which may be continuously drawn in the direction of diminishing λ until it reaches the piston (point P on the line $V = 1$). Therefore, if the piston velocity exceeds the value to which corresponds the point $P_\mathfrak{J}$ then a flow regime with an overdriven detonation wave of definite strength will establish. In this flow the gas particle velocity rises towards the piston.

If the Chapman-Jouget regime occurs (point $\mathfrak{J}$) along with the integral curve $\mathfrak{J}P_\mathfrak{J}$, which corresponds to a defi-nite piston velocity, on can move from the point $\mathfrak{J}$ in the direc-tion of diminishing λ along the integral curve $\mathfrak{J}A$. The veloci-ty of gas then diminishes and becomes zero at point A . This point A corresponds to a weak discontinuity with $\frac{r}{t} = a$. Passing this discontinuity the solution may be continued along the integ-ral curve $V = 0$, which corresponds to quiet gas, in the direc-tion of rising z ; the value of λ then tends to zero.

The solution constructed in this manner describes the gas flow behind the Ch.-J. wave propagating from a point source. Let us note that in accordance with Eq. (2.11b) in this solution $\frac{dV}{d\lambda} \to \infty$ immediately behind the Ch.-J. wave and $\frac{dV}{d\lambda} = 0$ at the outer side of the region where the gas is at rest.

The integral curve $\mathfrak{J}A$ lies in the region under the parabola (2.14), where $a + v < \frac{r}{t}$, therefore from an arbi-trary point of this curve one can jump through a shock wave into a point of that part of domain $OA\mathcal{E}$ which is situated above the parabola (2.14) (all these points constitute the dotted line ЪA on Fig. 4b) and then continue the solution continuously until it reaches the piston. These solutions correspond to the piston velocities in the range from the value corresponding to continuous flow behind Ch.-J. wave and down to zero.

Fig. 5 presents graphs of velocity distribution behind the detonation wave for different values of piston velocity.

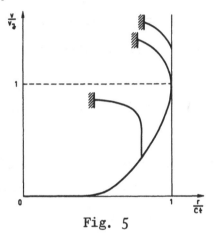

Fig. 5

We note that similar to the case of plane waves the solution of the problem with given piston velocity and fast combustion front propagating through the gas will not be unique. Beginning from any point of the curve OЪ which corresponds to a given value of heat release, one can move along the integral curve towards diminishing λ and then pass through a shock wave to another integral curve corresponding to compression flow continuating to piston.

In the examples considered the gas is initially at rest and is uniform, the regime of selfsustaining detonation which establishes after ignition is then necessarily the Ch.-J. regime. This statement gave the reason for often used qualitative speculations of following character. If the detonation

wave is overdriven $c < \upsilon + a$ the rarefaction waves, progressing through combustion products with sonic speed will overtake
the detonation wave, interact with it and diminish the speed of
wave propagation consequently the speed of the detonation wave
may be conserved only if $c = \upsilon + a$, i.e. only in the Ch.-J. regime.
However it has to be notified that in general the previous arguments are not correct. For instance when the undisturbed gas
is not uniform the selfsustaining detonation wave may be overdriven. In the case of converging cylindrical or spherical detonation waves the velocity of the wave is not constant and rises infinitely as the wave collapses.

1.3. ASYMPTOTIC BEHAVIOUR OF DECAYING DETONATION WAVES

In some cases of initiation of detonative combustion the detonation wave is initially strongly overdriven. This
will happen for instance, in the case of concentrated energy re
lease in combustible gas mixture. With the time the detonation
wave will get weaker under the influence of rarefaction waves
overtaking it from behind and can approach the limiting Ch.-J.
regime. Let us investigate the laws of its approach to this limit [3,4].

The values of the velocity, pressure and density
behind a detonation wave which propagates through a quiescent
gas can be expressed by Eqs. (2.4). Let us consider the waves
whose strength is only slightly in excess of the strength of Ch.-

J. wave. For these waves the quantity $\varepsilon = 1 - \dfrac{q}{q_z}$ is small and it is possible to obtain the following approximate representations for the flow parameters behind the wave (assuming that the wave is sufficiently strong, i.e. q_z is not close to unity) :

$$\frac{\upsilon}{\upsilon_z} = 1 + \sqrt{\frac{1+q_z}{1-q_z}}\,\varepsilon^{1/2} + \frac{1+q_z}{2(1-q_z)}\,\varepsilon + \dots ,$$

$$\frac{p}{p_z} = 1 + \gamma\,\frac{1-q_z}{\gamma+q_z}\left(\sqrt{\frac{1+q_z}{1-q_z}}\,\varepsilon^{1/2} + \frac{1}{1-q_z}\,\varepsilon + \dots\right),$$

$$\frac{\varrho_z}{\varrho} = 1 - \frac{1-q_z}{\gamma+q_z}\left(\sqrt{\frac{1+q_z}{1-q_z}}\,\varepsilon^{1/2} + \frac{q_z}{1-q_z}\,\varepsilon + \dots\right).$$

Here $\upsilon_z, p_z, \varrho_z$ are values of the velocity, pressure and density behind the Ch.-J. wave :

$$\upsilon_z = \frac{c_z}{\gamma+1}(1-q_z), \quad p_z = \varrho_z c_z^2\,\frac{\gamma+q_z}{\gamma(\gamma+1)}$$

$$\varrho_z = \varrho_1\,\frac{\gamma+1}{\gamma+q_z}.$$

It is easy to check that quantities $\dfrac{p}{\varrho^\gamma}$ and $a - \dfrac{\gamma-1}{2}\upsilon$ are constant up to terms of the order of $\varepsilon^{1/2}$, that is, with this degree of accuracy the gas parameters behind the wave satisfy relationships of a Riemann wave propagating through combustion product towards the detonation wave.

The equations of one-dimensional gas motion, together with conditions on the detonation wave, enable us to express the values of derivatives with respect to the coordinate r of gas dynamic functions behind the wave through gas parameters on the wave and the quantity $\dfrac{dq}{dr}$ which characterizes the acceleration of the wave. Thus, for the derivative $\dfrac{\partial v}{\partial r}\Big|_s$ we obtain the expression

$$\frac{\partial v}{\partial r}\Big|_s = \frac{2(1+q)+\sqrt{(1-qq_{\check{z}})\left(1-\dfrac{q}{q_{\check{z}}}\right)}}{2(1-qq_{\check{z}})\left(1-\dfrac{q}{q_{\check{z}}}\right)}\,\frac{v}{q}\,\frac{dq}{dr} - (\nu-1)\frac{qv}{r}\,\frac{1+\dfrac{\gamma v}{a_1\sqrt{q}}}{\sqrt{(1-qq_{\check{z}})\left(1-\dfrac{q}{q_{\check{z}}}\right)}}.$$

Let us first examine the planar wave of detonation. With an accuracy of $\varepsilon^{1/2}$ the flow behind it is a Riemann wave. For such a wave

$$v = \Phi(r-(a+v)t),$$

$$a - \frac{\gamma-1}{2}\,v = a_{\check{z}} - \frac{\gamma-1}{2}\,v_{\check{z}}$$

where Φ is an arbitrary function, the form of which determines the shape of the progressing wave. Let $\Phi(\xi)$ be such that $\Phi(\xi_0) = v_{\check{z}}$ and $r\,\Phi'(\xi)\to\infty$ as $r\to\infty$ and $\xi\to\xi_0$, where ξ_0 is the limiting value of $\xi = r-(a+v)t$ on the detonation wave.

Since

$$\frac{\partial v}{\partial r} = \frac{\Phi'(\xi)}{1+\dfrac{\gamma+1}{2}\,t\,\Phi'(\xi)}$$

and $c_\chi t \longrightarrow r_\delta$ for $r \longrightarrow \infty$,   then

$$\frac{\partial v}{\partial r} \longrightarrow \frac{2 c_\chi}{[(\gamma+1) r_\delta]} .$$

Comparing this expression with earlier written general expression for $\dfrac{\partial v}{\partial r}$ on the wave, we obtain the asymptotic relationship

$$\frac{2}{r_\delta} = -\frac{1}{\varepsilon} \frac{d\varepsilon}{dr_\delta} .$$

After integrating twice we find the asymptotic law of propagation of a planar detonation wave and the asymptotic expressions for the gas parameters behind the wave

$$c_\chi(t-t_0) = r_\delta\left(1 + \frac{r_0^2}{2 r_\delta^2} + ...\right)$$

$$\frac{v}{v_\chi} = 1 + \sqrt{\frac{1+q}{1-q}} \ \frac{r_0}{r} + ... ,$$

$$\frac{p}{p_\chi} = 1 + \gamma \frac{1-q}{\gamma+q} \sqrt{\frac{1+q}{1-q}} \ \frac{r_0}{r} + ... ,$$

$$\frac{\rho}{\rho_\chi} = 1 + \frac{1-q}{\gamma+q} \sqrt{\frac{1+q}{1-q}} \ \frac{r_0}{r} + ...$$

Here t_0 and r_0 are some constants, index χ by q is omitted.

In this manner the planar detonation wave on

weakening approaches the asymptote

$$r - c_{\mathfrak{z}}t = const$$

and the detonation regime asymptotically approaches the Ch.-J. regime. We note that the usual planar shock wave does not approach any asymptote on degeneration into an acoustic wave ; it crosses the straight line $r - a_1 t$ = const at an arbitrarily large value of the constant on the right-hand side of the equation of straight line.

In contrast to flows with planar waves, the transition to the Ch.-J. detonation regime in the propagation of cylindrical or spherical detonation waves through a quiescent gas can take place over a finite distance.

For small ε the expression for $\dfrac{\partial \upsilon}{\partial r}$ behind the wave has the form

$$r_{\mathfrak{z}} \frac{\partial \upsilon}{\partial r_{\mathfrak{z}}} = - \frac{\upsilon_{\mathfrak{z}}}{1-q_{\mathfrak{z}}} \frac{r_{\mathfrak{z}}}{\varepsilon} \frac{d\varepsilon}{dr_{\mathfrak{z}}} - (\nu-1) \frac{1 + \dfrac{\gamma\upsilon_{\mathfrak{z}}}{a_1\sqrt{q_{\mathfrak{z}}}}}{\sqrt{1-q_{\mathfrak{z}}^2}\ \varepsilon^{1/2}} \ .$$

For self-similar motions (ε = const) with $\nu \neq 1$, it follows from this that for $\varepsilon \to 0$ the quantity $r\dfrac{\partial \upsilon}{\partial r}$ is negative, and in absolute magnitude it tends to infinity like $\varepsilon^{-1/2}$. Let the flow behind the detonation wave weaken it in such a manner that for $\varepsilon \to 0$ the absolute value of $r\dfrac{\partial \upsilon}{\partial r}$ approaches infinity slower then in the case of self-similar motions.

Then

$$r_{\flat} \frac{d\varepsilon}{dr_{\flat}} = - N \varepsilon^{1/2}$$

where $N > 0$. From here we obtain after integration

$$\varepsilon^{1/2} = \varepsilon_0^{1/2} - \frac{N}{2} \ln \frac{r_{\flat}}{r_{\flat 0}}.$$

According to this formula $\varepsilon \to 0$ for finite $r_{\flat}$, that is, the transition to Ch.-J. detonation regime may take place over a finite distance.

Let us give a short description of the analysis of the flow structure near the point of transition of the detonation to the Ch.-J. regime.

In the t, r plane let the point O with coordinates t_0, $r_0 = c_{\flat} t_0$ be the point of transition of the overdriven detonation wave $\mathcal{D}O$ into the Ch.-J. wave $O\mathcal{J}$ (Fig.6). We will look for the solution of Eqs. (2.7) behind the detonation wave in the vicinity of the point O.

As it is well known, if on the line $\lambda = \lambda_0(\tau)$ the values of functions $\upsilon = \upsilon_0(\tau)$, $a = a_0(\tau)$, $\flat = \flat_0(\tau)$ do not satisfy the

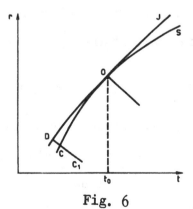

Fig. 6

characteristic relations (2.8) the solution of Eqs. (2.7) in
the vicinity of this line may be presented by series in the following form

(3.1) $$v - v_0 = \bar{v}_1 (\lambda - \lambda_0) + \bar{v}_2 (\lambda - \lambda_0)^2 + \ldots .$$

The successive coefficients of these series are uniquely determined through the values of functions on the line $\lambda = \lambda_0(\tau)$. If
the initial data do satisfy the characteristic relations then
the value of one of the first coefficients of series (3.1) at
some value of τ may be prescribed arbitrarily (for instance
$\bar{v}_1(0)$ for the solution with initial data on the acoustic characteristics, $\bar{\delta}_1(0)$ – for the solution with initial data on
the particle trajectory).

If the given functions satisfy only the first
of the two characteristic relations (2.8) the line $\lambda = \lambda_0(\tau)$
is then an envelope of characteristics.

The solution of the form (3.1) does not exist in
this case. But it may be proved that Eqs. (2.7) have solution
in the form of the following series

(3.2) $$v = v_0 + v_1 \sqrt{\lambda - \lambda_0} + v_2 (\lambda_0 - \lambda) + \ldots$$

and similar series for a and δ and that the coefficients of
these series are determined by initial values on the line $\lambda = \lambda_0(\tau)$.

Thus the series (3.2) give the solution of Eqs.
(2.7) depending on arbitrary functions λ_0, v_0, a_0, δ_0,

which are related by one of conditions (2.8a).

These series may be used to obtain the solution in the vicinity of the point where the overdriven spherical or cylindrical detonation wave transforms into Ch.-J. wave.

Let us consider first the flow behind the segment $O\dot{\tau}$ of the Ch.-J. wave.

This wave propagates with constant speed and thus the parameters of the gas behind it are constant ; as a consequence with $\nu \neq 1$ the second characteristic relation (2.8) is not satisfied in this case. Therefore the Ch.-J. wave with $\nu \neq 1$ is an envelope of acoustic characteristics. According to previous statement the solution for the quantities υ , p , ρ , behind it will have the form ($c_{\dot{\tau}}$ is taken for unity) :

$$\frac{\upsilon}{\upsilon_{\dot{\tau}}} = 1 + V_1 \sqrt{1-\lambda} + \ldots \qquad \frac{p}{p_{\dot{\tau}}} = 1 + p_1 \sqrt{1-\lambda} + \ldots$$

$$\frac{\rho}{\rho_{\dot{\tau}}} = 1 + R_1 \sqrt{1-\lambda} + \ldots$$

where

$$V_1 = \pm \sqrt{2 \frac{(\nu-1)(\gamma+q_{\dot{\tau}})}{(1-q_{\dot{\tau}})(\gamma+1)}} \qquad p_1 = \gamma R_1 = \gamma \frac{1-q_{\dot{\tau}}}{\gamma+q_{\dot{\tau}}} .$$

All next coefficients of series (3.3) will be evidently constant, so the flow behind the Ch.-J. wave with $\nu \neq 1$ (differently from the case $\nu = 1$) will be necessary self-similar - such as

if the detonation wave would be Ch.-J. wave all the time beginning from the moment $t = 0$. According to the possibility of
choosing different signs in the expression for V_1 two different self-similar flows with Ch.-J. detonation wave are possible.

These flows have been described in Sec. 1.2.

Behind the segment of the wave $\mathcal{D}0$, which pre-
ceeds the occurence of Ch.-J. regime, the flow is represented
again by the formulae (3.2) where the functions λ_0, a_0 , v_0 ,
ϑ_0 satisfy the condition (2.8a) with upper sign. The coefficients in the expansions of functions λ_0 , v_0 ,..., v_1 ,... in
powers of τ are expressed in a definite manner through the coef_
ficients in the equation of the detonation wave $\mathcal{D}0$

$$\lambda_0 = 1 + d_2 \tau^2 + d_3 \tau^3 + \ldots$$

The self-similar expansion wave behind the Ch.-J.
detonation wave $0\mathcal{J}$ is adjoined through a shock or a characteris_
tic $0S$ to the flow which can be connected in a continuous manner along the characteristic $C0$ with the flow behind the wave
$\mathcal{D}0$. This connecting flow may have either the same form (3.2)
as the flow in the region $\mathcal{D}0C$, but with different expressions
for λ_0, v_0 ,..., v_1 ,..., or it may not have a singularity at
the point 0 (in the case where the line $0S$ is a characteristic), that is it can be represented by expansions in integral
powers of $\lambda_0 - \lambda$.

The found solution shows the manner in which the transition of the overdriven detonation wave into a Ch.-J. wave takes place when the wave is weakened by disturbances approaching it from the rear. In particular, this solution describes the transition from the self-similar compression flow with a Ch.-J. wave to a self-similar rarefaction wave behind the same wave. Such transition arises when a spherical (or cylindrical) piston supporting the Ch.-J. wave with compression flow behind it suddenly stops (Fig. 7). The rarefaction disturbances coming from the piston transform the self-similar compression flow I into self-similar rarefaction flow II leaving the strength of the detonation wave unaltered.

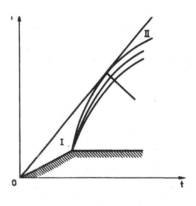

Fig. 7

The numerical solution of the strong explosion problem in combustible gas mixture with instantaneous heat release on the wave front confirms the conclusions of the general theory described above [5]. On the Fig.8 the dimensionless coordinates are shown where the spherical overdriven detonation wave generated by explosion transforms into Ch.-J. wave. These data show that this transformation occurs when the radius of the detonation wave is of the same order of magnitude as the radius at which the heat released by explosion equals to the initial blast energy. The same is true for cylindrical wave.

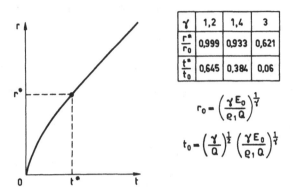

γ	1,2	1,4	3
$\frac{r^*}{r_0}$	0,999	0,933	0,621
$\frac{t^*}{t_0}$	0,645	0,384	0,06

$$r_0 = \left(\frac{\gamma E_0}{\varrho_1 Q}\right)^{\frac{1}{\gamma}}$$

$$t_0 = \left(\frac{\gamma}{Q}\right)^{\frac{1}{2}} \left(\frac{\gamma E_0}{\varrho_1 Q}\right)^{\frac{1}{\gamma}}$$

Fig. 8

Contrary the velocity of planar wave is still considerably in excess of the velocity of Ch.-J. wave even at much larger distances.

Thus if the heat release zone is infinitely thin the overdriven detonation wave produced, for instance, by expanding piston, or by concentrated energy release tranforms when the outer energy source stops to act into a self-supporting Ch.-J. detonation wave.

Therefore, the following statement is justified : the self-similar flow with Ch.-J. detonation wave propagating from an igniting source of infinitesimal size may be considered as asymptotic state of non self-similar flows which arise in combustible gas mixtures due to the expansion and following stopping of a piston or to a concentrated blast.

1.4. NONUNIQUENESS OF SOLUTIONS WITH INFINITELY THIN HEAT RELEASE FRONTS

The statement made at the end of the previous section requires the assumption that beginning from some moment of the initial stage of motion (particularly from the first moment of motion) and permanently later the combustion proceeds in a detonation wave.

Actually, along with the solutions of problems on piston expansion or on concentrated blast in a combustible mixture when gas burns in a detonation wave there are possible different solutions of same problems (under assumption on infinitely thin combustion zone). In these solutions the shock wave is not followed by instantaneous heat release, but the mixture burns in a slow combustion front which propagates through the gas with given velocity(which is usually small as compared with the sound velocity). If one assumes that the combustion permanently proceeds in a slow combustion front we may consider as asymptotic state of all these flows the self-similar solution of the problem of igniting source with propagating slow combustion front instead of detonation wave.

We will not follow the details of getting this solution and only the structure of it will be described.

In the case of planar waves the solution consists

of an adiabatic shock, uniform flow behind it, flame front and
a region with gas at rest between the flame front and the plane,
where ignition took place. If in the flame front the Ch.-J. regime is reached, the flame front and the quiet gas at the igniting plane are separated by a centered Riemann rarefaction wave.

In the case of spherical (or cylindrical) waves
the solution of the problem with igniting source consists of
successively situated shock wave, continuous compression wave,
flame front and a second compression region between flame front
and centre. If the combustion proceeds in Ch.-J. regime then bet
ween the flame front and the second compression wave an expansion wave forms, which terminates by a shock wave.

In this manner the asymptotic flow field on large
distances from the region where the ignition of the gas took
place may be different ; the gas may burn either in a detona-   -
tion wave or in a slow combustion front. In general, if there
are no particular limitations made, the solutions of problems on
the motion of combustible gas mixtures with instantaneous burning of a gas in a detonation wave or in a flame front with given
propagation velocity are determined not uniquely. It may be assumed that at any arbitrarily chosen instant the heat release in
a detonation wave stops and the combustion proceeds further in a
slow combustion front going out from the detonation wave. This
possibility is illustrated on Fig.9, where $\mathcal{D}O$ is the detonation wave, OF –flame front, OS –shock wave, OR –centred

Riemann wave, OL —contact surface. Obviously, the velocity of
the flame front cannot be assumed to be higher than a certain
limit.

A reverse possibility can occur : the flame
front at any instant may transform into detonation wave. This
transformation of flame front into Ch.—J. wave is illustrated on
Fig. 10.

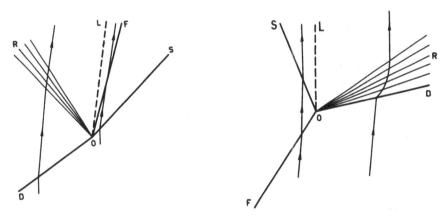

Figs. 9 and 10

Others, more complicated case are possible too, for instance,
that shown on Fig. 11. The gas in the disturbed region ahead of
the slow combustion front starts to burn in a detonation wave.
The origin point of detonation wave may be chosen arbitrarily.

Under real conditions of concentrated energy
release different flows may develop — with detonation wave or
with slow combustion front. Examples of flow are known in which
the detonation wave already formed splits into an adiabatic shock
and a flame front or, vice versa, the flame front changes into

detonation wave or a detonation wave forms in the flow ahead of
the flame front.

On Fig. 12 a bullet flying through hydrogen-air
mixture is shown.

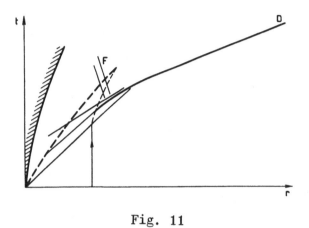

Fig. 11

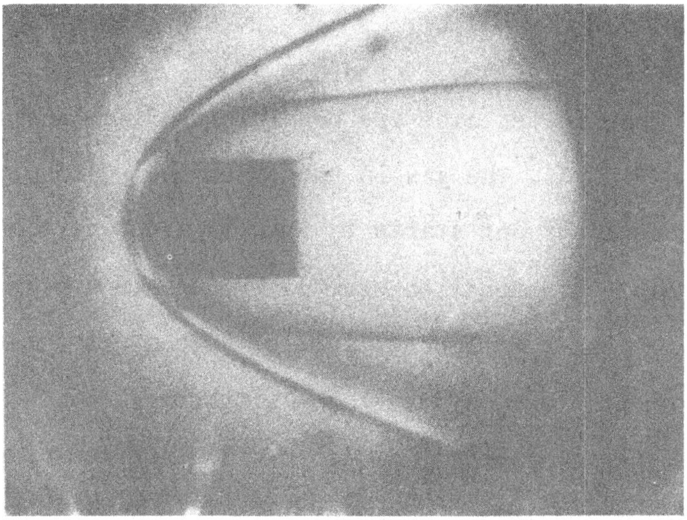

Fig. 12

The bullet acts here as a piston which causes initially an over driven detonation wave gradually decaying by action of the rarefaction disturbances coming from the piston in the course of its declaration. In this case the transition to the Ch.-J.regime with following burning of gas in the Ch.-J. detonation wave does not occur. Instead of this the detonation wave splits into adiabatic shock wave and slow combustion front.

Similarly on Fig. 13 the strong shock wave generated initially by concentrated energy release (focusing a laser beam) in the hydrogen-oxigen mixture does not transform into Ch.-J. detonation wave, but splits into adiabatic shock wave and flame front.

Fig. 14 shows the combustion of gas ignited along the axis of a cylinder. It can be seen that the combustion which initially proceeds in a slow combustion front turns into a detonation.

It is important to emphasize that in many cases of flows including different regimes of combustion, as in above examples, the thickness of the heat release zone and the linear scale of the region where the transition from one mode of combustion to another takes place, are very small as compared with characteristic linear dimension of the flow field. Therefore the dynamical analysis of these flows may be based on assumption on infinitely thin heat release fronts. However to exclude the arbitrariness stated above in the choosing of combustion

regimes additional conditions are needed and these conditions
are substantially related to the finite rates of chemical reac
tions which lead to the heat release. The finite chemical reaction rate introduces into consideration characteristic times or
lenghts.

t

t

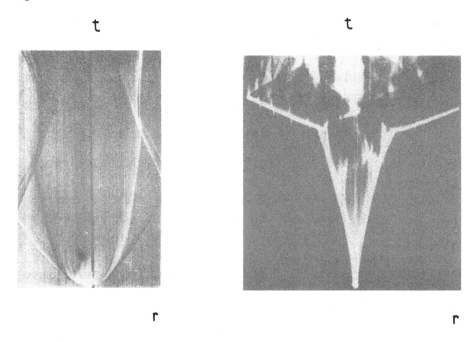

r

r

Figs. 13 and 14

During the early stage of combustion development they cannot be
neglected compared with scales of the whole flow field ; these
influence determines the flow development and, in particular,
its asymptotic structure at the time when the scales, characteristic for the chemical reactions, may already be considered as
vanishingly small. At these stages of the motion, when the
thickness of the heat release front may be neglected in the
scale of the flow field, the existence of a definite internal
structure of the front may lead when the local conditions change,

to such development of the process inside of the heat release
front and in an adjacent zone which dimension is comparable
with the thickness of the front which will change one combustion regime to the other.

In this connection it is obvious how important
it is to obtain solutions of problems on gas flows with heat
release accounting for finite chemical reactions rates in the
gas which has passed the shock wave.

1.5. MODELS OF DETONATION WAVES WITH INTERNAL STRUCTURE

The chemical reactions in gases have mostly chain
mechanism, i.e. in the course of reaction between initial substances many new intermediate species will be formed and these
species are responsible for the continuation of the reaction.
The combustion of gas mixtures is a complicated process consist
ing of many elementary reactions. The rates of these reactions
determine the rate of the global reaction. Only in a limited
number of simplest reactions their rates are known with sufficient accuracy, in most cases the rates are known very approximately or are unknown at all.

Therefore to treat theoretically the phenomena
in the chemical reactions zone, different simplifying assumptions on the structure of the detonation wave are used. In all
these models the flow in the combustion zone is assumed to be

one-dimensional.

1º) The simplest model accounting for finite
chemical reactions rates is the two-fronts model (suggested by
K.I. Shchelkin). This model considers the detonation wave as
consisting of a shock wave and following it flame front, the
position of the flame front being determined by induction time
or by combustion lag. The gas particle which has crossed the
shock wave moves adiabatically during a certain induction time
τ , after what the heat release in the particle and changing
of its chemical composition take place instantaneously.

The induction time is determined by a conditional
chemical reaction which proceeds in the particle

$$\frac{dc}{dt} = f(p,T)$$

c being equal to zero at shock wave and $c = 1$ at the combustion front. This gives for τ the following formula

(5.1) $$1 = \int_{t}^{t+\tau} f(p,T)\,dt$$

where t is the time when the particle crosses the shock wave.
The function $f(p,T)$ usually contains the Arrhenius multiplier

$$\exp\left(-\frac{E}{RT}\right)$$

1a – We can even consider a simpler model not accounting the influence on τ of the changes of flow parameters between the shock wave and the flame front, i.e. putting simply

$$\tau = \frac{1}{f(p_\delta, T_\delta)}$$

where p_δ, T_δ are pressure and temperature immediately after the shock wave.

2°) Another model of the detonation wave is the so-called $\mathfrak{D}NZ$ –model (named so by Döring, Neumann and Zeldowich). In this model the gas behind the shock wave is considered as consisting of a mixture of two conditional reacting species A and B. As a result of reaction the species A with adiabatic exponent γ and molecular weight μ transforms into combustion product B which is characterized by new values of γ and μ. The reaction $A \longrightarrow B$ is accompanied by heat release and begins immediately behind the shock wave according to the equation

$$\frac{d\beta}{dt} = -K\beta^m p^\ell e^{-\frac{E}{RT}}.$$

Here β is the mass concentration of the initial species A, m –order of reaction (for bimolecular $m = 2$), K, ℓ are constants (in most investigations is assumed that $\ell = m-1$), E –the activation energy.

2a – Some authors account for the reverse reac-

tion $B \longrightarrow A$; thus the kinetic equation in this case takes
the form

$$\frac{d\beta}{dt} = -K\beta^m p^\ell \exp\left(-\frac{E}{RT}\right) + K_1(1-\beta)^{m_1} p^{\ell_1} \exp\left(-\frac{E+Q_m}{RT}\right)$$

where m , ℓ , m_1 , ℓ_1 , K , K_1 are constants, Q_m is the maximal
heat which can be added to the gas in the course of reaction.
In this case the equilibrium value of β differs from zero and
depends on values of constants in the kinetic equation and on
values of thermodynamic parameters in the equilibrium state.

For real combustible mixtures $Q_m \gg E$ and the
reaction $A \rightleftarrows B$ proceeds mainly as a direct one ; in such cases
it is not needed to complicate the model by accounting for the
reverse reaction.

3°) One can use the model of detonation wave which
includes both models 1 and 2, i.e. to assume that the heat is
released not instantaneously, as it does in model 1, but is released in the course of reaction $A \rightleftarrows B$ proceeding with finite
rate, as in model 2, however this reaction does not start immediately behind the shock wave, but begins only after the induction time has gone. This model is relatively simple and at
the same time it enables to account for important characteristic features of different chemical reactions. The constants
entering the kinetic equations may be determined by experiments
or by computing one-dimensional steady flows of gas mixtures

with a sufficiently complete set of elementary reactions acting in the chain mechanism.

4°) Accounting for the complete set of elementary reactions for real combustible mixture is a very cumbersome problem. To simplify it we might use the fact that in most cases depending on conditions only one part of elementary reactions is substantial, but it is difficult to determine in advance which part of the reactions is important. Moreover, as was told earlier, the data concerning the rates of elementary chemical processes are known within accuracy of one-two orders of magnitude.

Let us consider more detailed the reaction of hydrogen burning. As for the characteristic conditions we will take the conditions in conventional shock tubes or ballistic ranges. The initial temperature and pressure in these installations are usually not high ($T_1 \sim 300°\,K$, $p_1 \sim 1$ atm or less), thus the combustion may proceed only after the shock wave has passed the gas. As example, if a bullet moves through the range with pressure $0,1 - 1$ atm and at room temperature with speed of $2 - 3$ km/sec then in the vicinity of the strongest part of the bow wave the pressure will be of the order of $5 - 15$ atm and the temperature rises to $2000 - 3500°\,K$. Under these conditions the reaction between components of the mixture proceeds extremely fast.

The main features of the hydrogen combustion are studied in sufficiently details. It is known that the chemical reactions develop by chain mechanism with formation of active radicals H, O and OH. These radicals are intermediate substances and their appearance leads to exponential growth of the total reaction rate with heat release and formation of the final product - water.

Two qualitatively different periods are characteristic for the reaction. The initial period is characterized by formation of chains, generation of active centres (radicals) without any considerable heat release. This period is the induction time. After the sufficient quantity of radicals has been generated, the rate of their formation fastens sharply, the "overproduction" of radicals takes place, i.e. they are generated to larger amount, than it would be in equilibrium state. The heat is released mainly as result of recombination of atomic species and formation of water. The second period is the period of heat release.

The question on the choice of the system of elementary reactions and of the values of constants characterizing the reaction rates for hydrogen-oxygen and hydrogen-air mixtures, though studied better than for other mixtures, is not finally solved and is discussed in a series of recent papers.

Usually it is assumed that the basic reacting components are H_2, O_2, O, H, OH and H_2O. These components characterize the main properties of the combustible mixture flows under broad variety of conditions. Along with these components in the course of chain reaction many other unstable and less active radicals are formed. Their importance is strongly dependent on the flow conditions and has not been studied suffi ciently yet. It concerns the species like HO_2, H_2O_2 or appear ing in the hydrogen-air mixture species NH_3, NH and others.

The choice of the kinetic model may be based on the results of the investigations of one-dimensional flows. Thus, for example the detailed analysis of the hydrogen-air flows in nozzles has shown that the presence of radicals with low activity does not influence considerably the flow parame- ters. However we cannot conclude from this result that under other conditions, for instance in the flow around body in which the compression of gas is followed by its expansion, the influ- ence would be the same.

Accounting for a series of different investiga- tions one can use the following kinetic model for hydrogen-air mixture.

The air is assumed to consist of oxygen and inert, non-dissociated nitrogen ; the oxygen may dissociate and react with hydrogen. In the flow six main components men- tioned above are formed andan inert component N_2 is present. The inert degrees of freedom are excitated in equilibrium manner.

The components interact by following eight reac- tions

$$H + O_2 = OH + O \qquad K_1 = K_2 = 3 \cdot 10^{14} e^{-8.81/\bar{T}} \qquad 1)$$

2) $O + H_2 = OH + H$ $K_1 = K_2 = 3 \cdot 10^{14} e^{-4,03/\bar{T}}$

3) $H_2 + OH = H + H_2O$ $K_1 = K_2 = 3 \cdot 10^{14} e^{-3,02/\bar{T}}$

4) $2\,OH = O + H_2O$ $K_1 = K_2 = 3 \cdot 10^{14} e^{-3,02/\bar{T}}$

5) $2H + M = H_2 + M$ $K_1 = 10^{16}, \quad K_2 = 10^{15}$

6) $OH + H + M = H_2O + M$ $K_1 = 10^{17}, \quad K_2 = 10^{16}$

7) $O + H + M = OH + M$ $K_1 = 10^{16}, \quad K_2 = 10^{15}$

8) $2O + M = O_2 + M$ $K_1 = 6 \cdot 10^{14}, \quad K_2 = 3 \cdot 10^{14}.$

Here K_1 , an K_2 denote the upper and lower limits of the experimental values of constants in chemical reactions rates, $\bar{T}$
is the absolute temperature in Kelvin's degrees, divided by 10^3,
the dimension of the constants for bimolecular reaction (1)–
(4) is cm^3 / mole.sec, that for trimolecular reactions (5)–(8)
is cm^6 /mole.sec. M denotes any molecule or atom which is considered as third body ;   the concentration of the third body
is equal to the sum of concentrations of all species, including
N_2 .

          5°) A more complicate version of the model 4 ac-
counts for the presence of law-activity radicals HO_2 and H_2O_2

by adding to the system of reactions (1)-(8) the following four

$$H_2 + O_2 = 2OH \qquad\qquad 9)$$

$$HO_2 + M = H + O_2 + M \qquad\qquad 10)$$

$$H_2 + HO_2 = H_2O_2 + H \qquad\qquad 11)$$

$$H_2O + HO_2 = H_2O_2 + OH . \qquad\qquad 12)$$

The constants of reverse reactions have been taken in the form

$$K_r = A_i \, T^n \, exp\left(\frac{-E_i}{RT}\right).$$

For the models under consideration the formulation of the problem assumes the shock wave to be infinitely thin and the boundary condition on the shock requires the equality of concentrations for all components ahead of the shock and behind of it. However the internal state of the molecules may change in non-equilibrium manner due to the differences in the excitation rates for translational, rotational and vibrational energies. Particularly under certain conditions the vibrational energy of a particle crossing the shock may not achieve its equilibrium value and therefore its value immediately behind the shock has to be taken equal to its undisturbed value (which may be assumed equal to zero).

In the model 5 for oxygen-air mixture the account has been taken for the non-equilibrium excitation of vibrations of H_2 , O_2 and N_2 molecules. The system of chemical kinetics equations is then completed by relaxational equations of the form

$$\frac{de_{vi}}{dt} = \frac{e_{vo}(T) - e_{vi}}{\tau} = \omega_{ev}(p,T)$$

where $e_{vo}(T)$ is the equilibrium value of vibrational energy, $\tau(p, T)$ – the relaxation time. Relaxation times for different collision processes of O_2 , H_2 and N_2 molecules are given by simple approximating expressions of the form

$$p\tau = 3,9 \cdot 10^{-10} exp\left(100\ T^{-1/3}\right) atm.\ sec \quad (for\ H_2 - H_2)$$

suggested by S.A. Losev.

2.1. INTRODUCTION AND FORMULATION OF THE PROBLEM

Until recently the numerical and analytical investigations of the steady combustion of gas mixtures at supersonic speeds were carried out under the assumption of one-dimensional flow behind a normal shock wave or quasi-one-dimensional
flow in a stream tube with given cross-section variations or
with given variations of one flow parameter (for instance, at
constant pressure).

For different applications the investigation of
two dimensional flows in internal channels or around bodies is
of considerable interest. In particular, the need for explanation of experimental results obtained in ballistic ranges led
to development of methods for calculations of two-dimensional
supersonic flow of combustible gas mixtures around bodies.

In the present lectures the problems are considered of two-dimensional flows around bodies with attached or
detached shock waves. Short explanation is given of basic data
concerning numerical methods used for different kinetic models
of combustion behind shock waves and some results of computations and their comparison with experimental evidence are discussed.

The system of differential equations governing
steady flow of reacting gas mixture expresses the conservation

laws of mass, momentum, energy and gas components and has the following form

$$\operatorname{div} \rho \bar{V} = 0 \qquad\qquad \rho \frac{d\bar{V}}{dt} + \operatorname{grad} p = 0$$

(1.1)
$$\frac{d}{dt}\left(\frac{V^2}{2} + h\right) = 0$$

$$\frac{d\alpha_i}{dt} = \omega_i \quad i = 1,\ldots,N \qquad \frac{de_{v_j}}{dt} = \omega_{v_j} \quad j = 1,\ldots,M.$$

Here $\bar{V}$ is the velocity vector, p –pressure, ρ –density, α_i –mass concentration of i –th component, ω_i –the rate of its production, N –the total number of reacting components, M –the total number of vibrationally excited molecules of different types. The number of kinetic differential equations may be reduced by the use of finite relations expressing the conservation of elements.

The system (1.1) must be completed by the expression for the total enthalpy of the mixture and the equation of state

(1.2) $\quad h = \sum\limits_{i=1}^{N} \alpha_i h_i , \quad p = \frac{\rho R T}{\mu} , \quad \frac{1}{\mu} = \sum\limits_{i=1}^{N} \frac{\alpha_i}{\mu_i} .$

The system of differential equations (1.1) together with finite relations (1.2) is closed . For two-dimensional problems the number of differential equations equal to $4 + N + M$.

To get the solution of equations the conserva-
tion laws across the shock wave and the condition on body surface should be used. These conditions have the form

at the shock wave

$$\rho_\infty V_{n\infty} = \rho_s V_{ns} \qquad V_{\tau\infty} = V_{\tau s}$$

$$\rho_\infty V_{n\infty}^2 + P_\infty = \rho_s V_{ns}^2 + P_s \qquad \frac{V_{n\infty}^2}{2} + h_\infty = \frac{V_{ns}^2}{2} + h_s$$

$$\alpha_{i\infty} = \alpha_{is} \tag{1.3}$$

$$e_{vj\infty} = e_{vjs}$$

on the body surface

$$V_n = 0 . \tag{1.4}$$

In the case of symmetric flow around a body of
revolution or symmetric profile the condition on the symmetry
line should be used.

In the problem of flow past a blunt-nosed body
the system of gasdynamic and kinetic equations is ellyptic in
the subsonic region and hyperbolic — in the supersonic. It is
expedient to describe separately the numerical methods used for
computations in subsonic and supersonic regions.

2.2. COMPUTATION OF SUBSONIC FLOW REGION

For numerical solutions of problems on steady supersonic flow around blunt-nosed bodies in USSR the method suggested by G.F. Telenin [6] and usually called the method of straight lines is widely used. In a series of papers [7-12] this method has been developed for application to combustible gas mixture flows. For the model of infinitely thin detonation wave the method of straight lines applies completely analogous to adiabatic gas flows, the only difference being that in the energy conservation law across the shock the heat release term Q appears.

The system of equation (1.1) simplifies in the case of infinitely thin detonation wave : the kinetic equations drop out and the completing relations reduce to one expression for the enthalpy of the perfect gas

$$(2.1) \qquad\qquad h = \frac{\gamma}{\gamma-1} \frac{p}{\rho}.$$

Let us exemplify the method of straight lines by the consideration of gas flow around a body of revolution with two-fronts model of detonation wave [8]. For the enthalpy we use the expression (2.1). In this case we only need one kinetic equation

$$\frac{dc}{dt} = f(p,T)$$

which determines the function c when the flow field is known.

Let us introduce the polar coordinates r, Θ and denote by υ, u the velocity components in radial and circumferential directions. On Fig. 1 the shock wave contour is denoted by $\mathfrak{s}$, the combustion front – by f , the body surface – by b . We introduce the nondimensional variables referring the linear dimensions to the bodies radius of curvature R_b at the stagnation point, the velocities – to the maximal speed V_{max} of the oncoming stream, the density – to ρ_∞ , the pressure – to $\rho_\infty V_{max}^2$. We write the system (1.1) for the variables u , υ , p and the stream function ψ :

$$\left[\gamma p^{\frac{\gamma-1}{\gamma}}\vartheta(\psi)-\upsilon^2\right]\frac{\partial \upsilon}{\partial r} - u\upsilon\frac{\partial u}{\partial r} + \frac{1}{r}\left\{\left[\gamma p^{\frac{\gamma-1}{\gamma}}\vartheta(\psi)-u^2\right]\frac{\partial u}{\partial r} - \right.$$

$$\left. - u\upsilon\frac{\partial \upsilon}{\partial \Theta}\right\} + \frac{\gamma p^{\frac{\gamma-1}{\gamma}}\vartheta(\psi)}{r}(2\upsilon+u\,\mathrm{ctg}\,\Theta) = 0$$

$$(2.2a)$$

$$\frac{\partial u}{\partial r} = \frac{1}{r}\frac{\partial \upsilon}{\partial \Theta} - \frac{u}{r} - \frac{\gamma}{\gamma-1}r\sin\Theta\frac{p}{\vartheta(\psi)}\frac{d\vartheta(\psi)}{d\psi}$$

$$\frac{\partial p}{\partial r} = \frac{p^{\frac{1}{\gamma}}}{\vartheta(\psi)} \left(\frac{u^2}{r} - \frac{u}{r} \frac{\partial v}{\partial \Theta} - v \frac{\partial v}{\partial r} \right)$$

$$\frac{\partial \psi}{\partial r} = \frac{1}{\vartheta(\psi)} p^{\frac{1}{\gamma}} r u \sin \Theta , \qquad v \frac{\partial c}{\partial r} + \frac{u}{r} \frac{\partial c}{\partial \Theta} = \tilde{f}(p, \psi)$$

(2.2b)
$$\vartheta(\psi) = \frac{p^{\frac{1}{\gamma}}}{\varrho} .$$

We transform the region 1 between the shock wave and the combustion front and the region 2 between the combustion front and the body surface into strips with unity width by introducing new independent variables $\xi^{(1)}$, $\xi^{(2)}$, and Θ according to formulae

$$\xi^{(1)} = \frac{r - r_f}{\varepsilon^{(1)}} \qquad \varepsilon^{(1)}(\Theta) = r_3 - r_f$$

$$\xi^{(2)} = \frac{r - r_b}{\varepsilon^{(2)}} \qquad \varepsilon^{(2)}(\Theta) = r_f - r_b$$

$$\Theta = \Theta .$$

We transform Eqs. (2.2) to these new variables and solve the equations with respect to derivatives on ξ . In each region we get then the equations of the form

(2.3)
$$\frac{\partial f_k}{\partial \xi} = F_k \left(\xi, \Theta, f_{\dot{\delta}}, \frac{\partial f_{\dot{\delta}}}{\partial \Theta} \right) \qquad \begin{array}{l} \dot{\delta} = 1, \ldots, 5 \\[2mm] K = 1, \ldots, 5 . \end{array}$$

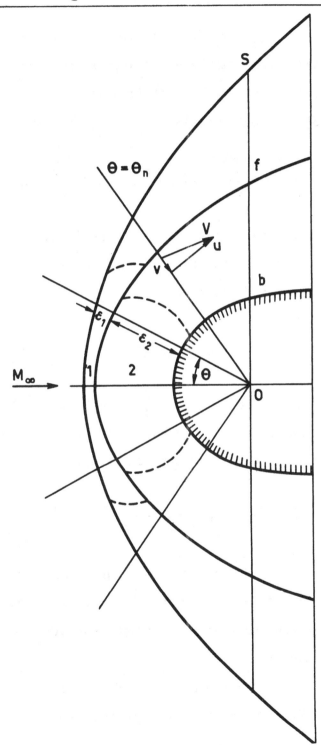

Fig. 1

After drawing $2n + 1$ rays $\Theta = $ const in the upper and lower halfplanes we approximate the derivatives on Θ using the values of functions at $2n + 1$ points by the interpolating polynomials

$$(2.4) \qquad f_\ell = \sum_{j=0}^{n} f_{\ell j}(\xi) \Theta^{2j}, \qquad f_m = \sum_{j=0}^{n} f_{m j}(\xi) \Theta^{2j+1}$$

for even and odd functions respectively. The equations of the shock wave and of the combustion front should be presented in the polynomial form as well

$$r_\vartheta = \sum_{j=0}^{n} r_{\vartheta j} \Theta^{2j} \qquad\qquad r_f = \sum_{j=0}^{n} r_{f j} \Theta^{2j}.$$

We replace the derivatives $\dfrac{\partial f_j}{\partial \Theta}$ in Eqs. (2.3) by their appro- ximate expressions. The conditions for these equations to be sa- tisfied along each of the rays yield the system

$$(2.5) \qquad \frac{df_{\kappa\ell}}{d\xi} = F_{\kappa\ell}(\xi, f_{ij}) \qquad\qquad \begin{array}{l} K, i = 1, \ldots, 5 \\[4pt] \ell, j = 0, 1, \ldots, n \end{array}.$$

We note that the equations in the system (2.5) for u and ψ on the symmetry line are satisfied identically ; therefore to rise the accuracy the equation for $u_0 = \dfrac{\partial u}{\partial \Theta}\bigg|_{\Theta=0}$ is usual ly introduced.

By the procedure described the boundary problem for equations in partial derivatives reduces to the boundary problem for $5n-1$ ordinary differential equations (2.5) for the same number of unknown functions.

      The numerical algorithm for solving the problem is the following. We prescribe $2n+2$ values of free parameters r_{sj}, r_{fj} ($j=0,1,...,n$) which determine the equations of the shock wave and of the combustion front. The conditions on the adiabatic shock permit to calculate the values of gasdynamical functions behind the shock waves at the node points. The boundary condition for c is $c=0$ at $r=r_s$ ($\xi^{(1)}=1$). With the initial data obtained the Cauchy problem for the system of Eqs. (2.5) should be solved up to the combustion front ($\xi^{(1)}=0$) using the explicit difference Runge–Kutta scheme. At the line $\xi^{(1)}=0$ the needed information is remembered, in particular the values c_m at the node points $\Theta=\Theta_m$ and the coefficients of the interpolating polynomials for the unknown functions as well. Then using the conditions on the combustion front which include the amount Q of the heat released we compute the initial data for the region 2 on the line $\xi^{(2)}=1$ and then solve the Cauchy problem up to the body contour $\xi^{(2)}=0$. Since initially free parameters were prescribed arbitrarily $2n+2$ conditions $c_m=1$ at $\xi^{(1)}=0$ and $v_m-\dfrac{r_b'}{r_b}u_m=0$ at $\xi^{(2)}=0$ $(m=0,1,...,n)$ will not be satisfied. The adjustement of the parameters proceeds according to the Newtonian scheme and we iterate until the conditions are satisfied with a given degree of accuracy.

      When the model 3^0 is used (see Sec.1.5)one non-

equilibrium reaction $A \rightarrow B$ is taken into account in the region
2. According to the general procedure of the method applied the
approximating ordinary differential equations should be integrat
ed in this case as well along the coordinate rays by the explicite Runge-Kutta difference scheme. However this procedure leads
to difficulties. In practically interesting cases a large part
of the region 2 is occupied by nearly equilibrium flow. When the
flow is close to the equilibrium conditions the right-hand side
term of the kinetic equation consists of the product of a small
multiplier (β^m) and a large quantity (Kp^ℓ). Consequently
small errors during the numerical integration of the system and
the computation of the gasdynamic functions may lead to large
errors in the evaluation of derivatives. The use of explicite
schemes with automatically changing step size has the disadvan-
tage that under flow conditions near equilibrium the steps become very small and sometimes even become computers zero.

When a flow with complicate kinetics is to be computed, additional difficulties arise due to difficulties of the
dividing beforehand the flow field into two zones – the induction zone and that of combustion, thus the system of equations
should be integrated across the whole flow field passing the regions with sharp and in some cases nonmonotone variations of
functions. As an example of complicate flow structure the Fig.2
shows the variation of the derivatives $\dfrac{d\alpha_i}{d\xi}$, $\dfrac{de_{vi}}{d\xi}$ on the
axis of symmetry ahead of sphere for one of the cases considered.

In the recent time a number of approaches has been suggested to overpass the difficulties mentioned above. It is well known that the explicit methods like the Runge–Kutta method are based on the expansion of the right-hand term in the equation $\frac{dy}{dx} = \varphi(x,y)$ into series on $\Delta x = x - x_1$:

$$\frac{dy}{dx} = A_0 + A_1 \Delta x + A_2 (\Delta x)^2 + \ldots$$

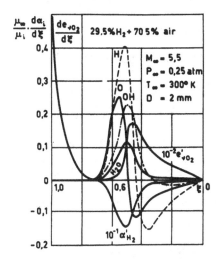

Fig. 2

If the gradients of the unknown function are large, the coefficients A_0 , A_1 , will be large and to integrate the equation with high accuracy one needs to use small steps Δx, or to make use of methods with higher order of expansion of the function $\varphi(x,y)$ The use of explicite methods for one-dimensional flows under conditions close to equilibrium turned out to be rather difficult, and for two-dimensional flows around bodies – practically impossible due to the exponential grow of the rounding errors.

The way to overcome these difficulties is the application of implicite schemes. The use of these schemes for the computation of one-dimensional nonequilibrium flows turned

out to be quite satisfactory. However in the case of two–dimensional flows the simple replacement of the explicite Runge–Kutta
scheme by implicite Eulerian scheme with iterations has not been
successful.

In this connection in the paper [9] a modification of the method of paper [6] has been   suggested which is
based on a numerical scheme similar to that for computation òf
one–dimensional flows with sharply varying functions.

In this modification from the node point of the
next layer an element of the streamline should be drawn back,
then a certain part of equations – for velocity components and
for pressure – should be integrated along the coordinate ray
$\Theta = \Theta_i$ according to explicite scheme, and another part – those
for density and for concentrations – along the streamline accor
ding to an implicite scheme. The effectiveness of this approach
becomes particularly evident in the case of flows with a compli
cate kinetic model 4° which we use in what follows to illustrate
the method [10].

The approximating system of equations with ξ and
Θ as independent variables has the form :
along the coordinate ray

$$(2.6a) \qquad \frac{d\upsilon}{d\xi} = \frac{\varepsilon}{\Delta} \left\{ \Phi_2 + \frac{\Phi_1}{Ar} \left[\frac{(a^2 - u^2)r'}{r} + u\upsilon \right] \right\}$$

$$\frac{du}{d\xi} = \frac{\varepsilon}{Ar} \Phi_1 - \frac{dv}{d\xi} \frac{r'}{r} \quad , \qquad \frac{dp}{d\xi} = \varrho \left(\frac{\varepsilon u^2}{r} - \frac{\varepsilon u}{r} v' - A \frac{dv}{d\xi} \right)$$

(2.6b)

along the stream line

$$\frac{d\Theta_1}{d\xi} = \frac{\varepsilon u}{Ar} \quad , \qquad \frac{d\ln\varrho}{d\xi} = \frac{1}{\gamma} \frac{d\ln p}{d\xi} - xX$$

(2.7)

$$\frac{dy_i}{d\xi} = x \frac{\omega_i}{\mu_i} + \frac{d\ln\varrho}{d\xi} y_i \qquad i = 1,...,4 .$$

In addition along the symmetry line

$$\frac{du_0}{d\xi} = - \frac{dv_0}{d\xi} \frac{r_0''}{r_0} + \frac{\varepsilon_0}{A_0 r_0} \left[\frac{p_0''}{\varrho_0} + u_0 (u_0 + v_0) \right]$$

(2.8)

where Θ_1 denotes Θ along the stream line

$$A = v - \frac{r'}{r} u \quad , \qquad a^2 = \frac{\gamma p}{\varrho} \quad , \qquad x = \frac{R_b \varepsilon}{V_{max} A}$$

$$X = \sum_{i=1}^{6} \left(\frac{\mu}{\mu_i} - \frac{h}{c_p T} \right) \frac{\omega_i}{\varrho} \quad , \qquad c_p = \sum_{i=1}^{7} c_{pi} \alpha_i$$

(2.9)

$$y_i = \varrho \frac{\alpha_i}{\mu_i} \quad , \qquad \gamma = \sum_{i=1}^{7} c_{pi} \alpha_i \Big/ \sum_{i=1}^{7} c_{vi} \alpha_i$$

Φ_1 , Φ_2 , Δ – are known function of u , v , p , ϱ , y_i .

Here the number of differential equations for concentrations y_i

is reduced to four by using the three finite relations expressing conservation of elements O , H and N .

       An elementary computational cell on the n-th ray
is shown on Fig. 3.

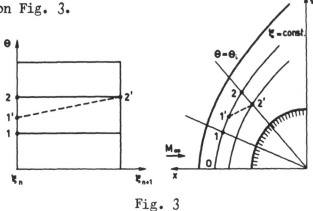

Fig. 3

The computations should be made in the following sequence. With
known values of functions in node points $\Theta = \Theta_i$ on the n-th
layer we compute the values of u , v , p , u_0 in the node
point on the ($n + 1$)-th layer using Eqs. (2.6) and (2.8).
Then using this values on the ($n + 1$)-th layer we compute the
density ρ and concentrations y_i in the node point on the
($n + 1$)-th layer (point 2') transforming the Eqs. (2.7) in
accordance with implicite Eulerian scheme

$$f_K^{n+1} = f_K^n + \Delta s \left[\delta F_K^n + (\delta - 1) F_K^{n+1} \right] \qquad 0 < \delta < 1 .$$

To determine the values of functions at the point 1' the polynomials (2.4) should be used.

      For the most part it has proved to be possible
to avoid iterations when determining the cross point of the

stream line with previous layer, i.e. it is possible to find Θ_1
by explicite formula. Then the determination of parameters on
the ($n+1$)-th layer is reduced to the solution of five tran-
scendental algebraic equations which proceeds by the Newton method [11].

The main shortcomings of these two schemes
are the following.

a) The necessity to iterate in each node point
and to make multiple matrix reversals when solving the system of
equations by Newton's method. These reversals increase the time
needed for computation.

b) The system of Eqs (2.6) – (2.8) has
singularities of type $\dfrac{0}{0}$ at points where $A = 0$, that is
where the velocity component normal to the line $\xi = \text{const}$
becomes zero. This singularity may appear inside the shock layer;
it is always present at the body surface. In the case of equilibrium flows the singularity does not appear in Eqs (2.5) due
to a special choice of variables (as one of them the stream func
tion is used). For nonequilibrium flows the attempts failed to
avoid the appearence of the singularity by the choice of variables. Therefore in the course of integration of system (2.6) –
(2.8) towards the body surface the unknown functions should
be extrapolated beginning from a certain distance from this surface.

In this connection the computation of the flow

around the forder part of a sphere in the case of hydrogen—oxygen and hydrogen—air mixtures with complicate kinetic model has been carried out by another modification of the numerical method of straight line [12, 13]

It is well known that the following equation may serve as the model equation for a relaxation process

(2.10)
$$\frac{dy}{dx} = \frac{y_e - y}{\tau}.$$

This form of equation determines the exponential behaviour of its solution y. In order to obtain analogous behaviour of the numerical solution the approximating equation should be taken in the form

$$\frac{dy}{dx} = \frac{1}{\tau_n}(y_n - y) + A_0 + A_1(x - x_n) + A_2(x - x_n)^2 + \ldots$$

In the paper [14] the general matrix presentation of the approximating equations for the system

$$\frac{dy_i}{dx} = \varphi_i(x, y_k)$$

is suggested in the following form

(2.11)
$$\frac{d\bar{y}}{dx} = -\vec{\vec{P}}(\bar{y} - \bar{y}_n) + \bar{A}_0 + \bar{A}_1(x - x_n) + \ldots$$

where the matrix $\vec{\vec{P}}$ is the generalization of the relaxation time in Eq. (2.10). There are several methods based on approximation (2.11) which differ by the choice of matrix $\vec{\vec{P}}$, the order of approximation in ($x - x_n$) and the method of solving the system (2.11).

The idea of the paper $\begin{bmatrix} 13 \end{bmatrix}$ is as follows. The system of Eqs. (2.2) is written with ξ and Θ as independent variables ; along the coordinate rays it takes the form

$$\frac{du}{d\xi} = \frac{1}{A} f_u \, , \quad \frac{dv}{d\xi} = \frac{1}{A} f_v$$

$$\frac{dp}{d\xi} = \frac{1}{A} f_p \, , \quad \frac{d\varrho}{d\xi} = \frac{1}{A} f_\varrho \, , \quad \frac{d\gamma_i}{d\xi} = \frac{1}{A} f_{\gamma_i} \, .$$

This system is equivalent to the following one

$$\frac{du}{d\xi} = \left(1 + \frac{Ak}{\psi_0} \right) \frac{du}{d\xi} - \frac{k}{\psi_0} f_u \, ,$$

where $\dfrac{k}{\psi_0}$ is the normalized multiplier. The derivatives on the right-hand side of equations are replaced through the difference terms

$$\frac{du}{d\xi} = \left(1 + \frac{Ak}{\psi_0} \right) \frac{u_{n+1}^s - u_n}{\Delta\xi} - \frac{k}{\psi_0} f_u \qquad (2.12)$$

and the system (2.12) is solved by the following iteration procedure. The initial values of u_{n+1}^0 , v_{n+1}^0 ,.... at the (n+1)-th point are given by the linear or parabolic extrapolation.

Then the system (2.12) is integrated on the

segment $\left[\xi_n, \xi_{n+1}\right]$ by the usual explicite Runge–Kutta method

or by Eulerian method. This gives new values for u_{n+1}^{1}, v_{n+1}^{1}, ...

The iterations continue until the conditions $\left|u_{n+1}^{s}-u_{n+1}^{s-1}\right|<\varepsilon$, ...

will be satisfied.

In this approach the matrix of transition to the

($n+1$)-th layer is diagonal, thus its elements $p_{ii}=-\left(1+A\dfrac{K}{\psi_0}\right)/\Delta\xi$

can always be made positive by the appropriate choice of the norma

lizing multiplier $\dfrac{K}{\psi_0}$. This reduces the rounding errors when

integrating on the segment. The use of the Runge–Kutta method

for each step is equivalent to retaining in the series presen-

tation of the right-hand side of equations of the terms includ-

ing fifth order term in ($x - x_n$).

The relaxation method of solving the system des-

cribed above turned out to be more convenient than the matrix

method.

When solving the system (2.12) there is no

need to compute the parameters on the body surface by extrapo-

lation because the indefiniteness $f_v = 0$, $A = 0$ eliminates

automatically. This makes it possible to compute regimes of flow

with inflammation of gas near the body surface, and in general

case it makes it easier to determine the shock wave position.

In the paper $\left[15\right]$ the computation of the flow

of hydrogen–oxygen mixture with model kinetics IV ahead of a

sphere is made by the inverse method (with shock shape given the

body shape should be determined) of "series truncation" suggested by M. Van Dyke.

Let us give a short account for the procedure used in this paper.

The stream function ψ and the parameter $\delta = \dfrac{d\ln\rho}{dt}$ are introduced. The gas dynamic equation for functions ψ, p, T and δ and six equations for concentrations y_i are written in polar coordinates r, Θ. The use of parameter δ instead of ρ is caused by the more smooth behaviour of the former as compared with the latter.

According to the "series truncation" method the unknown functions are presented in the series form in $\sin\Theta$ and $\cos\Theta$ (as in the paper [16]):

$$\psi(r,\Theta) = (\sin\Theta)^\nu \left[\psi_1(r) + \psi_2(r)\sin^2\Theta + 0(\sin^4\Theta)\right]$$

$$p(r,\Theta) = p_1(r)\cos^2\Theta + p_2(r)\sin^2\Theta + 0(\sin^4\Theta)$$

$$T(r,\Theta) = T_1(r)\cos^2\Theta + T_2(r)\sin^2\Theta + 0(\sin^4\Theta) \qquad (2.13)$$

$$\delta(r,\Theta) = \delta_1(r)\cos^2\Theta + \delta_2(r)\sin^2\Theta + 0(\sin^4\Theta)$$

$$y_i(r,\Theta) = y_i(r)\cos^2\Theta + z_i(r)\sin^2\Theta + 0(\sin^4\Theta)$$

where $\nu = 1,2$ for plane and axisymmetric flows respectively. These expressions are substituted in the basic system of equations and the terms with the same power of $\sin\Theta$ are grouped together resulting in the following set of equations

$$N_{K1} + N_{K2} \sin^2 \Theta + \ldots = 0 \qquad K = 1,\ldots,7 .$$

Here use is made of finite relations – equation of state, energy integral and conservation conditions for elements 0 and H .

The expressions N_{Km} $(k=1,\ldots,7,\ m=1,2\ldots)$ are linear ordinary differential operators.

The solution of the problem to the first approximation is found from the equations $N_{K1} = 0$, to the second approximation – from the equations $N_{K1} = 0$, $N_{K2} = 0$, etc. Each approximation set of equations contains the pressure term from the next approximation (e.g. N_{21} contains $p_2(r)$). To close the system the authors use additional assumption : $p_2 = 0$, if the first approximation is considered, $p_3 = 0$ if the second, etc.

To get the initial values behind the shock wave the expressions (2.13) are inserted into the shock wave relations and the same procedure is employed as for the derivation of approximating equations. The Cauchy's problem with initial data obtained is solved by the Runge–Kutta method. To the first approximation the system of seven and to the second approximation – of fourteen differential equations should be solved. The body contour is determined by the condition $\psi_1 = 0$ to the first approximation and by the condition $\psi_1 = 0$, $\psi_2 = 0$ – to the second.

In this method one also has to deal with difficulties when computing the flow regions with large gradients of
parameters or with nearly equilibrium conditions.

The authors use the linearization of kinetic
equations on each step of integration suggested by G. Moretti
[17] and improved in the paper [18]. On each step of integration the following assumptions are made : the constants of chemical reaction rates are determined by the temperature at the
n-th point and remain constant in the interval $[r_n, r_{n+1}]$; b)
parameter δ is constant ; c) the concentration Y of the third
body is constant. These assumptions simplify the linearization
of the kinetic equations.

The first order approximation system has the
following form

$$\psi_1' = q_1$$

$$p_1' = \left(\frac{\nu \psi_1}{r} - \frac{\varrho_1 r^\nu \delta_1}{\nu} - q_1 \right) \frac{\nu^2 \psi_1}{\varrho_1 r^{2\nu}} \gamma_s M_s^2$$

$$q_1' = \left[\frac{q_1^2}{\nu} - \left(\frac{2-\nu}{r} \psi_1 + \frac{\varrho_1 r^\nu \delta_1}{\nu} \right) q_1 - \frac{2 \varrho_1 r^{2(\nu-1)}}{\nu} \frac{p_1}{\gamma_s M_s^2} \right] \frac{1}{\psi_1}$$

$$\frac{1}{\varrho_1} \left[\sum_i \mu_i y_i h_i (T_1) + \frac{\nu^2 \psi_1^2}{2 \varrho_1^2 r^{2\nu}} \right] = H \qquad (2.14)$$

$$y_i' = - \frac{\varrho_1 r^\nu}{\nu \psi_1} \left(\sum_k A_{ik} y_k + B_i \right) \qquad i = 1, \ldots, 4.$$

Here γ_s , M_s are respectively adiabatic exponent and Mach number immediately behind the shock wave. The matrix A_{ik} and B_i are constant on each step of integration.

The sequence of computations on each step is the following. A certain value of parameter δ_1 is fixed and the three first equations (2.14) are integrated by the explicit Runge–Kutta scheme. The values of functions at the $(n+1)$ –th point thus obtained, the value δ_1 and the relation

$$(2.15) \qquad \left(\frac{\varrho_1}{\psi_1}\right)^{n+1} = \left(\frac{\varrho_1}{\psi_1}\right)^{n}\left(1+\frac{\Delta r}{u_1^0}\,\delta_1\right)$$

following from the definition of δ , constitute the closed system for determination of concentrations y_i at the $(n+1)$ – th point. The system of four linear differential equations with constant coefficients is solved approximately as in [18] .

The solution of the system

$$y_i' = \sum_{j=1}^{4} a_{ij}\, y_j + b_i \qquad i = 1,\dots,4$$

is taken in the polynomial form

$$y_i = \sum_{k=0}^{2} d_{ik} t^k$$

The requirement to the residual terms

$$R_i(t) = \sum_{k=1}^{2}\left(d_{ik}\, k\, t^{k-1} - \sum_{j=1}^{4} a_{ij}\, d_{jk}\, t^k\right) - \sum_{j=1}^{4} a_{ij}\, d_{j0} - b_i$$

to satisfy the following integral relations

$$\int_{0}^{\Delta t/2} R_i(t)\,dt = 0 \qquad \int_{\Delta t/2}^{\Delta t} R_i(t)\,dt = 0$$

gives eight algebraic linear systems for determination of eight
sets of coefficients d_{ik}. These systems are solved by triangulat-
ing the matrix of coefficients at unknown quantities.

Then one finds $Y = \sum_{i=1}^{6} y_i$, $T_1 = \dfrac{p_1}{Y}$ and from
the forth equation (2.14), which is the energy equation, ϱ_1 is
determined. There upon the relation (2.15) gives a new value
of δ_1 . The iterations follow until the condition $\left| \delta_1^k - \delta_1^{k-1} \right| < \varepsilon$
is satisfied.

Analogous but more complicate procedure holds
for the second approximation. The comparison of the first and
second approximations enables to make some conclusions concerning the convergence of the method.

Let us note that the range of admissible application of the second approximation is limited to relatively small
values of Θ . Therefore the "series truncation" method with
small number of terms retained does not permit to continue the
flow computation into the supersonic region.

2.3. COMPUTATION OF THE SUPERSONIC FLOW REGION

For numerical computations of two-dimensional steady supersonic flows past bodies the method of characteristics is widely used. In the papers dealing with flows with combustion, different variants of calculations have been suggested with transition from one strip to another and with drawing the characteristics in backward direction.

Let us exemplify the inverse method of characteristics by considering the flow of combustible mixture around body with formation of infinitely thin detonation wave $\begin{bmatrix} 7, & 19 \end{bmatrix}$. The equations of characteristics and the compatibility relations along them are written in the Euler's variables : along the characteristics of the 1 and 2 family

$$(3.1) \qquad \frac{dr}{dx} = \frac{BA \pm B}{\beta B \pm A} \, , \quad \frac{4}{1 + \delta^2} \, d\delta \pm \frac{\beta}{\varrho \lambda^2} \, dP \, \frac{A}{r(A \pm B)} \, dr = 0$$

along the stream lines

$$(3.2) \qquad \frac{dr}{dx} = tg \, \vartheta = \frac{A}{B} \, , \quad \frac{d\lambda^2}{2} + \frac{dP}{\varrho} = 0$$

where

$$(3.3) \qquad A = \delta \left| 1 - \delta^2 \right|, \quad B = \frac{1}{4} \left(1 - \delta^2 \right) - \delta^2, \quad \delta = tg \, \vartheta / 4, \quad \beta = \sqrt{H^2 - 1}.$$

Here $\lambda = \dfrac{V}{a_*}$ reduced velocity, ϑ – the angle of the velocity vector with x – axis.

The computation of the supersonic flow region starts from the initial data on the limiting ray $\Theta = \Theta_n$ obtained in the course of the computation of subsonic region I (Fig.4) by the straight lines method.

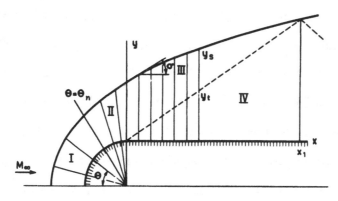

Fig. 4

In the paper [7] the flow around body of revolution or profile with discontinuity of contour curvature at $\Theta = \dfrac{\pi}{2}$ has been investigated. In the computation downstream from the section $\Theta = \dfrac{\pi}{2}$ the coordinates r, x, and in the region II where $\Theta_n \leq \Theta \leq \dfrac{\pi}{2}$ polar coordinates have been used.

The region between the shock wave and the body surface is transformed like in Sec.2.1 into the strip $[0,1]$ by introduction of the variable ξ.

Let us explain the numerical procedure. Let the solution be known on the line $x = x_n$ at the node points $\xi = \xi_i$ ($i = 0,..k-1, k, k+1...K$). We look for the solution on the next layer

$x = x_n + \Delta x$ at the same values of $\xi = \xi_i$. The computa-

tion starts from a point 3 at the shock wave (Fig. 5a) by adjust-

ing the value of $tg\,\delta$ (δ – the angle which the shock wave

makes with x direction) to meet the condition (3.1) along the

characteristics of the first family 1–3.

When computing the point 3 inside of the shock

layer (Fig. 5b) the values of unknown functions at this point

are taken in advance approximately (e.g. the same as at the

corresponding point on the layer x_n). Then the characteris-

tics of both families 1–3, 2–3 and the streamline 4–3 are drawn

from this point backwards.With known values of the functions at

the points 1,2,4 on the layer x_n we correct the values at the

point 3 using the relations (3.1), (3.2), etc. To compute the

flow at the point 3 on the body contour (Fig. 5c) the differ-

ence equations along the characteristics of the second family

should be used.

To obtain the values at the points of intersec-

tion of characteristics with the line x_n the through interpo-

lation of the unknown functions based on the values of these

functions at all node points of this line is used

$$f_i = \sum_{j=0}^{m} a_{ij}\, \xi^{j}.$$

The size of the step Δx should be chosen de-

pendently on the value of $\Delta\xi$ in order to meet the needed requirements on the accuracy and stability of the computational

scheme.

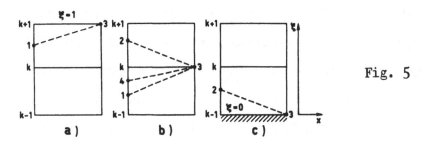

Fig. 5

If the body contour is smooth (sphere) the whole

supersonic region can be computed by a simple algorithm. In the

examples considered in $[7]$ the computations were carried out

consecutively in the following regions (Fig. 4) : region II

$\Theta_n \leqslant \Theta \leqslant \dfrac{\pi}{2}$, $0 \leqslant \xi \leqslant 1$; region III $0 \leqslant x \leqslant x_1, y_t \leqslant y \leqslant y_s$

region IV $0 \leqslant x \leqslant x_1$, $y_b \leqslant y \leqslant y_t$. By y_b, y_s are, respectively,

denoted the ordinates of the body contour and of the shock wave,

y_t is the ordinate of the characteristic emitting from the

point on the body contour where the curvature is discontinuous,

x_1 —the abscisse of the intersection point of this characteris-

tic with the shock. The boundaries of each region are normali-

zed by introducing the variable ξ , and the interpolation is

carried out by using the values at the points belonging to this

region only.

In the papers $[20, 21]$ the inverse method of cha-

racteristics has been used to compute the supersonic part of

the flow around blunted cones and the flow around sharp wedge

and sharp cone with attached shock wave. The kinetic model 2a
with one reversible reaction has been assumed. The computation
with this model differs from the described above by additional
kinetic terms in the relations along the characteristics and
stream lines, while across the shock wave the usual Rankine-
Hugoniot conditions hold. Besides that, the quadratic interpola
tion over three neighbouring points has been used on the n-th layer.

   A more general scheme of the three-dimensional
method of characteristics has been used in papers [22, 23] for
computation of two- and three-dimensional flows around pointed
bodies. The kinetic model with one non-reversible exothermic reaction in the heating zone has been assumed in this computation.

   The initial system of gasdynamic and kinetic equa
tions written in cylindrical coordinates (r, x, φ) has been trans
formed to new variables (ξ, x, φ), where

$$\xi = \frac{r - r_b(x, \psi)}{r_s(x, \psi) - r_b(x, \psi)}.$$

   This system of three-dimensional differential
equations has been reduced to the approximating two-dimensional
system by eliminating the derivatives with respect to φ. To do
this in the region $0 \leqslant \varphi \leqslant \pi$ one draws $\ell + 1$ meridional equidis-
tant in φ planes $\varphi = \varphi_K = \frac{k\pi}{\ell}$ ($k_, = 0,1...\ell$). The functions to be
differentiated by φ are represented by interpolating trigono-
metric polynomials with interpolation nodes in the planes $\varphi = \varphi_k$:

the odd functions in the form

$$F(\xi, x, \varphi) = \sum_{k=1}^{l-1} a_k(\xi, x) \sin k\varphi \qquad (3.4a)$$

the even – in the form

$$\Phi(\xi, x, \varphi) = \sum_{x=0}^{l} b_k(\xi, x) \cos k\varphi \qquad (3.4b)$$

where a_k, b_k are determined by the value of functions on the meridional planes. After substituting the derivatives $(\partial F / \partial \varphi)_k$ $(\partial \Phi / \partial \varphi)_k$ obtained from (3.4) into the transformed system of initial equations one gets the approximating system of two–dimensional differential equations with partial derivatives with ξ and x as independent variables. The dependent variables are the values of the functions to be found on the planes $\varphi = \varphi_k$.

The approximating system has two families of characteristics. One more family of curves possesses the characteristic properties similarly to the case of axisymmetric flows, these curves may by denoted as stream lines.

After this the inverse scheme of the method of characteristics for two–dimensional case is used, and the approximating system is solved numerically in a series of node points in all meridional planes $\varphi = \varphi_k$ simultaneously. The computing procedure is quite similar to that previously described.

In conclusion to this section let us emphasize certain advantages of the inverse method of characteristics as compared with the direct one : a) the coordinates of the node

points are known in advance ; b) it is easy to change the computation mesh ; c) the values of the functions are evaluated in
the sections x = const at fixed points what is convenient when
using the results of computation ; d) the scheme is of the second order of accuracy (if the iteration are applied) what enables to spare the computing time by choosing larger steps in x
direction with given requirements to the accuracy ; e) the conditions imposed by the stability criterium are less rigid then
in the case of the direct method of characteristics.

The disadvantages of the method are the following:
a) the influence domain of the initial differential equations is
taken into account only approximately ;
b) it is difficult to determine with sufficient accuracy the generation of internal shocks, though these shocks arise in many
flows with chemical reactions.

2.4. SOME RESULTS OF COMPUTATIONS

Let us consider some interesting results of computations giving the idea about the main properties of the super
sonic flow of combustible gas mixture around bodies of different
shape.

Fig. 6 presents the results for the flow around
a sphere-cylinder (Fig. 6a) and a flat plate with rounded nose
(Fig. 6b) with infinitely thin detonation wave. The influence
of the heat release parameter q is shown on the position of the
detonation wave, of the sonic line and of the characteristics
with discontinuous derivatives of flow parameters (dashed lines).

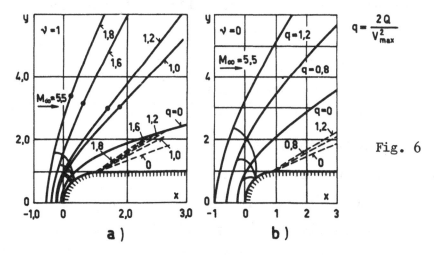

Fig. 6

With increasing q the stand-off distance of the detonation
wave from the body increases. Similarly to one-dimensional non-
steady flows (see Sec. 1.3)   the numerical results show [24]

that the strong detonation wave, which forms ahead of a blunt-
nosed axisymmetric body, gradually weakens and transforms into
Chapman-Jouget wave at finite distance from the body. The transi
tion points are shown on Fig. 6a by black circles. In the case
of plane flows the Chapman-Jouget regime has not been reached
within the limits of the computed flow domain.

However, as was previously told, the experimental
evidence [24-26] shows that in many cases the weakening detonation wave does not transform into Chapman-Jouget wave but at
some distance from the body it splits into the adiabatic shock
wave and the slow combustion front (see Pt. I, Fig. 12 for the
photograph*) of a bullet moving through the stoichiometric hydrogen-air mixture with velocity   2,1 Km/sec and $p_\infty = 0,5$ atm,
$T_\infty = 293°$ K). Using the kinetic models 1° and 3° this phenome-
non can be well understood and described quantitatively. With
the model 1° the product $p_\infty R_b$ is the similarity parameter. The
influence of this parameter on the position of shock wave,   combustion front and sonic line is shown on Fig. 7. It is seen
that when $p_\infty R_b > 10$ atm.mm the mixture inside of the region considered burns in a vanishingly thin zone so that the flow is in
equilibrium practically everywhere ; of course it does not follow from here that the detonation takes place also at large dis-

---

*) The photograph is made in the ballistic range of the Insti-
   tute of Mechanics of Moscow University.

tances from the body. When $p_\infty R_b < 0{,}01$ atm. mm the flow is frozen, that is the gas starts to burn in the immediate vicinity of the sphere surface.

Let us notify the discontinuity of the sonic line on the combustion front in this model.

With T_∞ fixed the Mach number M_∞ of the on-coming stream strongly influences the flow pattern. When M_∞ varies the stand-off distance of the shock wave from the stagnation point ε_0 varies nonmonotonically: within a certain range of M_∞ the shock wave moves upstream as the Mach number increases (Fig. 8).

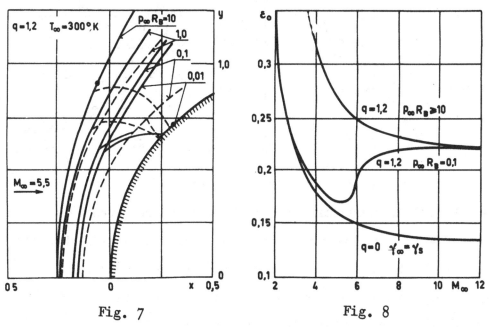

Fig. 7   Fig. 8

In the model 3° with one non-reversible reaction behind the induction zone two similarity parameters appear :

$p_\infty R_b$ and Γ_p / R_b.

       Fig. 9a shows for two different values of Γ_p the flow around a sphere–cylinder, and the Fig. 9b gives the density and pressure ditributions across the shock layer through different sections normal to the body surface. One can see that the detonation wave splits, and the combustion front emitted from the wave tends rapidly to a streamline separating the combustion products from the unburnt mixture.

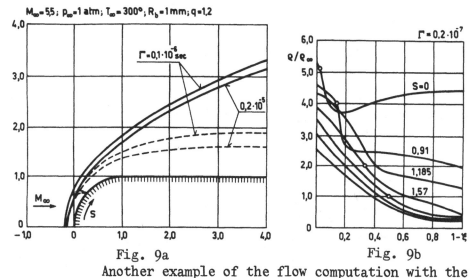

Fig. 9a   Fig. 9b

       Another example of the flow computation with the same kinetic model behind the shock wave is presented by the Fig. 10 when the flow pattern is shown for the wedge with halfangle equal to 30° [21] . With the conditions considered, the infinitely thin detonation wave would be overdriven. As the characteristic length L the initial width of the induction zone is taken, the dashed line on the Fig. 10a marking the rear boundary of this zone. As the density distributions across different

sections show (Fig. 10b) a compression wave propagates through
the flow due to the heat deposition. The compression wave reflects from the shock wave and reaches the wedge surface. The
density of the gas varies along the wedge surface in nonmonoton-
ic way and oscillates with decreasing amplitude (Fig. 10c).

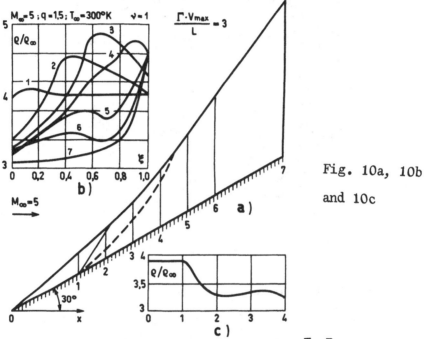

Fig. 10a, 10b
and 10c

This confirms the results of the linear theory [27].

On the Fig. 11 and Fig. 12 the results of computations are presented for axisymmetric flows around a cone with
halfangle equal to 30° and around a ogive-nosed cylinder [23].
The model 3a with reversible reaction behind the induction
zone has been used in these computations. Qualitatively the flow
past a cone is similar to that past a wedge. The strength of
the shock wave increases with the distance from the leading

point and the width of the induction zone decreases.

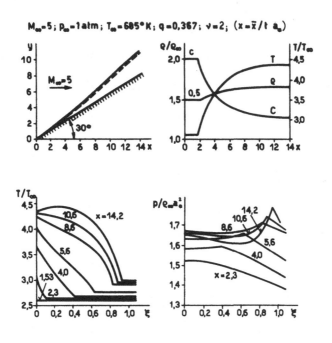

Fig. 11

Due to the heat release an internal shock forms in the flow
field ; however it is difficult to detect this shock using the
inverse scheme of the method of characteristics. The pressure
and the temperature rise along the cone surface, the pressure
distribution having a maximum. It is interesting to notify that
at $x < 12$ the maximum of the temperature through cross section
is on the cone surface, and when approaching to the equilibrium
conditions (at $x > 12$) this maximum moves inside of the shock
layer. The pressure at a given section rises in the induction
zone when moving from the shock wave and has a peak at the com-

bustion front, behind this front the rarefaction occur, the pressure having a minimum at some distance from the apex.

The flow past the ogive-cone is quite similar to
the flow past sphere-cylinder considered earlier. Initially the
induction zone in this flow is very narrow but then the splitting
of the detonation wave occurs. The pressure gradually diminishes
along the body surface whereas the temperature distribution along it has a maximum. Across the shock layer the temperature
rises gradually towards the body.

The calculations of flows around the body with detailed kinetic model 4° [10] allowed to investigate more precisely the structure of the combustion zone behind the detached
shock wave.

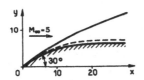

$$M_\infty = 5; \; p_\infty = 1 \, atm; \; T_\infty = 685°K; \; q = 0,367; \; \nu = 2; \; (x = \bar{x}/t \, a_*)$$

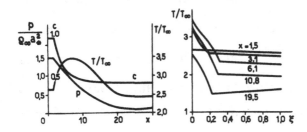

Fig. 12

Fig. 13 shows in the case of hydrogen–air flow around a sphere

the influence of the pressure p_∞ , the excess oxidant ratio α

and the diameter $\mathfrak{D}$ of the sphere on the distributions of reac-

ting components, velocity and the temperature of the gas and on

the flow pattern (the dashed lines on Fig.13b and 13c correspond

to the adiabatic flow). With decreasing pressure the width of

the induction zone and of the heat release zone broaden, the for

mer being less sensitive to pressure variation than the latter.

The increasing diameter of the sphere leads to the diminution of

the relative thickness of both these zones. Let us notify that

for a non–equilibrium flow with only bimolecular reactions the

binary similarity holds, the similarity parameter being $p_\infty \mathfrak{D}$.

Therefore in the induction zone where the most significant re-

actions are 1) – 4) (pp. 53–54) the appropriate similarity has

place, whereas in the heat release zone this similarity violates.

If the mixture becomes leaner, the concentration of atomic hydro

gen at a fixed point decreases, whereas content of water changes

non–monotonically. Its maximum is reached for the stoichiometric

mixture ($\alpha = 1$). The thickness of the shock layer has its maximum

at $\alpha = 1$ as well. The gas accelerates along the axis of symmetry

and its velocity behind the combustion front may be in excess of

the velocity immediately behind the shock wave (Fig. 13c).

The more complex model 5° made it possible to in-

vestigate the influence on the flow of the vibrational relaxa-

tion of molecules of O_2 , H_2 and N_2 , and of the presence of the

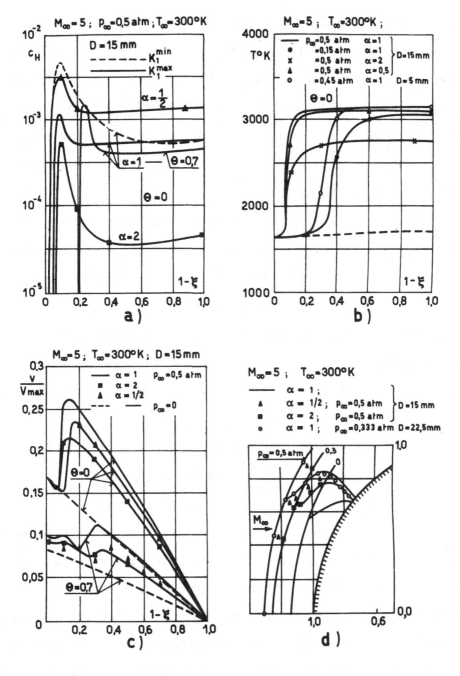

Fig. 13

slightly active radicals HO_2 and H_2O_2 [12].

At moderate speeds ($V_\infty \sim 2\,Km/sec$ and $p_\infty \sim 0.1\,atm$) of the hydrogen-air mixture these radicals have only small influence on the flow parameters around the body. But under the same conditions in the mixture $2O_2 + H_2$ the account for the presence of HO_2 and H_2O_2 leads to a considerable rise of the equilibrium temperature.

The behaviour of the gasdynamic parameters u , v , p , ϱ and T is only slightly dependent on the composition of the mixture being substantially determined by the value of the heat release. These parameters behave similarly to the case of the solutions of model problems with the same amount of heat released.

With increasing velocity of the oncoming stream not only the position of the heat release zone changes, but the amount of the heat deposition decreases (Fig. 14 a). As a result of this at the velocities corresponding to $M_\infty \sim 10$ the mixture does not burn in the subsonic region ahead of the body.

At small values of the binary similarity parameter $p_\infty \mathcal{D}$ the vibrational relaxation may play an important rule. The temperature of the gas immediately behind the shock wave with equilibrium excitation of the vibrational energy is lower than with frozen vibrational energy (Fig. 14b). In the induction zone the equilibrium temperature (solid line) does not change notably whereas the non-equilibrium temperature (dotted line) de-

creases due to the partial absorption of the kinetic energy by
the vibrational excitation below the equilibrium value. Since
during the earlier stage of the induction time the temperature
had stronger influence on its duration, the account for the non-
equilibrium excitation leads in general to the induction time re
duction. Simultaneously the period of the heat deposition increases, thus partially compensating the influence of the vibra
tional relaxation on the thickness of chemical reactions zone.

Fig. 14c and 14d show the influence of the pressure
p_∞ on the distribution of the concentrations H , OH and H_2O
along the symmetry line and the distribution of Mach number for
$2O_2 + H_2$ mixture. Let us notify the sharp maximum of atomic
hydrogen concentration; more smooth maximum possesses the concentration of hydroxyl OH. On Fig.15 some computational results
taken from [15] are presented on the flow of hydrogen-oxygen mix
ture. The curves represent the first approximation solution. The
graphs on Fig. 15a show the influence of the shock wave curvature on its stand-off distance from the body and on the ratio of
the pressure of the stagnation point to its value for frozen
flow. The distributions of the reacting components and gasdynamic parameters between the shock wave and the body surface are
shown on Fig. 15b and Fig. 15c. Qualitatively these results a-
gree with what described above. The thickness of the heat release zone on Fig. 15b is larger than under similar conditions,
to which the Fig. 13 and the Fig. 14 refer.

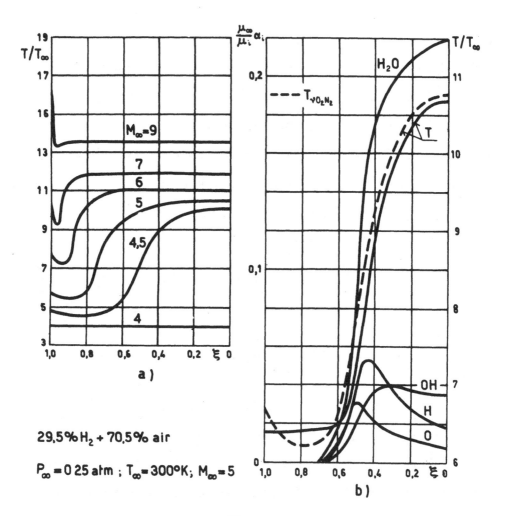

29,5% H_2 + 70,5% air

P_∞ = 0 25 atm ; T_∞ = 300°K; M_∞ = 5

Fig. 14

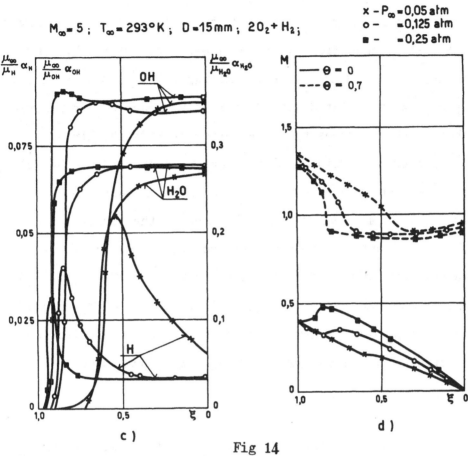

Fig 14

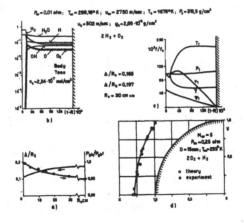

Fig. 15

This discrepancy may be explained partially by certain differen

ce of initial conditions in both cases, particularly by lower

pressure $p_\infty = 0,01 \, atm$ taken in the paper $[15]$. Besides the dis

crepancy may be related with different accuracy of the numerical

methods used. It seems that the method of straight lines is rath

er in preference. This conclusion may be supported by the follow

ing comparison with an experiment under conditions close to equi

librium. The numerical results obtained by the straight lines

methods $[12]$ and experimental evidence $[22]$ give for the rela-

tive stand-off distance of the shock wave from the sphere, appro

ximately the same value $\varepsilon \sim 0,27$. On the other hand, according

to paper $[15]$ the maximum value of ε for nearly equilibrium

flow even a reacher mixture ($2H_2 + O_2$) is approximately 0,194 on

ly.

The variation in the course of computation of the

rates of individual reactions or either neglecting some of them

or some of reacting components enable us to make certain conclu-

sions concerning the importance of different processes and the

deposit of elementary reactions into production of different com

ponents.

Under distinct conditions the importance of diffe

rent reactions may vary. However it is stated that for the $2O_2 + H_2$

mixture in all cases the most significant reaction is the 3)-th

(for convenience the reactions are renumerated on the Fig. 16).

This is illustrated by the Fig. 16a, which gives the production rates of atomic hydrogen in each particular reaction and the total production rate of this component (the solid line). The role of the reaction (3) in production of other components is similar. A noticeable effect may be caused by the reactions (4.6), (4.7) and (4.8) as well.

It must be notified that for the stoichoimetric hydrogen–oxygen mixture the variation of reaction rates by an order of parameters does not change considerably the gasdynamic parameters (Fig. 16b) and the width of the shock layer (Fig.16c). However a careful comparison with the experiment presented in Fig. 16c seems to show the preferableness of using the constants which lie close to the lower limits of their experimental values.

The influence of the changing by one order of magnitude of the reaction rates is more noticeable for the $2O_2+$ $+H_2$ mixture. The results of the corresponding investigation are presented on the Fig. 16d. If one decreases the rate of reaction 3) by one order of magnitude or neglect this reaction the heat release ceases almost entirely.

It follows from above that the comparison of the computation results with experiments in a broad range of external conditions permit to make useful conclusions concerning the kinetics of the hydrogen–oxygen reaction.

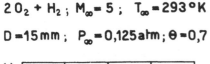

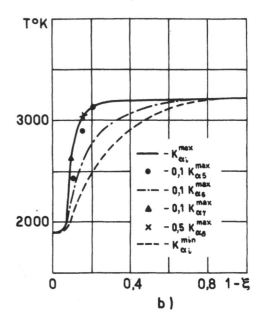

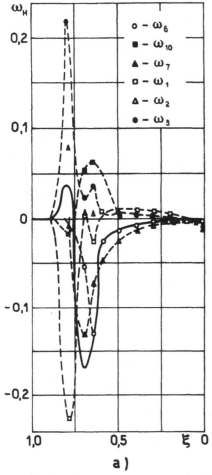

b)

a)

1) $H + O_2 = OH + O$

2) $O + H_2 = OH + H$

3) $H_2 + OH = H + H_2O$

4) $2OH = O + H_2O$

5) $2H + M = H_2 + M$

6) $OH + H + M = H_2O + M$

7) $O + H + M = OH + M$

8) $2O + M = O_2 + M$

9) $H_2 + O_2 = 2OH$

10) $HO_2 + M = H + O_2 + M$

11) $H_2 + HO_2 = H_2O_2 + H$

12) $H_2O + HO_2 = H_2O_2 + OH$

Fig. 16

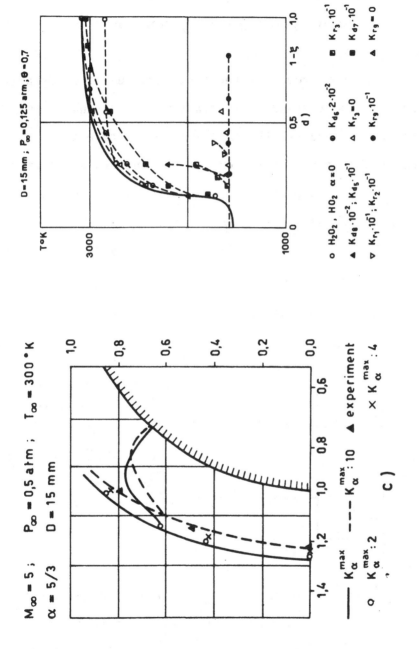

Fig. 16

3.1. INSTABILITY OF THE FLAME FRONT IN A
TWO—FRONT DETONATION WAVE

Let us consider the distributions of gasdynamical parameters which correspond to a steady planar detonation wave with a two-front structure (Fig. 1) and let us take these distributions as the initial state of the gas on both sides of the flame front. As it was previously told the use of the conservation laws on the flame front only is unsufficient to determine the gas motion which arises from the initial state under consideration; it is necessary to prescribe the velocity of flame propagation through the unburnt gas.

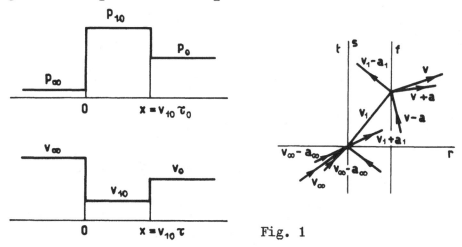

Fig. 1

Therefore in the problem of flow development from the given initial state with discontinuous distributions of gasdynamical quantities the velocity of the flame front serves as a parameter whose value is to be determined by the additional con-

dition of its correspondence to the ignition time. Let us analyse the variety of all flows which can arise from the given in
itial discontinuous distributions with different values of the
velocity of flame front propagation (we denote all these flows
as gasdynamically possible). From the initial discontinuity
through the unburnt gas propagates the flame front preceded by
either a shock wave or by a centred rarefaction wave, or through
the unburnt mixture a detonation wave propagates. Through the
combustion products a shock wave or a centred rarefaction wave
propagates. All these possibilities have been analyzed in the
paper by G.M. Bam-Zelikovich [28] , see also book [29] .

In the case under consideration the initial distributions obviously give the solution of the problem with the
velocity c_f of the flame propagation through the gas which is
equal to the velocity of the unburned gas υ_{10} (in the system of
coordinates related to the initial flame front position, the velocity of the flame front $c'_f = \upsilon_{10} - c_f$ in this case is equal to
zero).

Let us analyse following the paper [30] , the
dependence of the flame front velocity upon the pressure p'_1 a-
head of it (behind the shock wave or rarefaction wave). The flow
with a shock wave preceding the flame front is schematically
shown on Fig. 2. The increase of the shock strength is related
to the rising of the flame front velocity (Fig. 3). At some value of this velocity $c_f = c_{f\mathfrak{z}}$ the relative velocity of the gas

behind the flame front will be equal to the sound speed (it may

happen just in the initial state) and at larger values of c_f the

flame front is accompanied by a centred rarefaction wave. As the

value of c_f continues to rise the flame front approaches the

shock wave until at some $c_f = c_{dz}$ they merge forming a normal de-

tonation wave. The further increase of c_f, i.e. the formation of

an overdriven detonation wave, is impossible in the analyzed

case. As a matter of fact behind such a wave the pressure of the

gas rises and the velocity diminishes as compared with the cor-

responding values in the unburned mixture. On the other side,

the initial pressure of the burnt gas is lower and its velocity

higher than these values in the unburnt gas

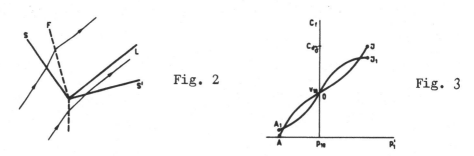

Fig. 2   Fig. 3

Therefore, it is evident that it is impossible

to make equal the pressures and the velocities on both sides of

the contact surface dividing the combustion products behind the

detonation wave and behind the flame front, neither with a shock

wave nor with a rarefaction wave propagating through burnt gas.

Let us consider now the flows in which the flame

front is preceded by a rarefaction wave. This case is illustrat-

ed on Fig. 4. The increase of the rarefaction wave intensity
(falling p_4') is related to the diminution of c_f (Fig. 3). The
intensity of the wave reaches its maximum
when c_f equals to zero, what means that
the combustion stops, i.e. that the initial discontinuity initiate the flow in
an inert gas.

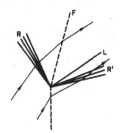

Fig. 4

With known thermodynam-
ic properties of the unburnt and burnt gases the curve AOJ on
Fig.3 is completely determined by the values of parameters of
the unburnt gas and by the amount of the heat released on the
flame front.

Among all gasdynamically possible solutions of
the problem with initial discontinuity only those may be realized in which the velocity of the flame front corresponds to the
induction time as determined by formula (5.1) of Pt. I, where
the integral is evaluated along the particle path in the disturb
ed flow.

Now we will determine that value of the veloci-
ty $c_{f_{chem}}$ of the flame front through an unburnt gas behind the
shock wave or the rarefaction wave which is in accordance with
the condition of induction time sparing by the particle.(Let us
notify that in the case of a rarefaction wave $c_{f_{chem}}$ will be constant only if the function $f(p,T)$ does not contain constants
enabling to form of determining data scales of time or length).

The shape of the curve determining the dependence of $c_{f\,chem}$ on p_1' evidently depends on the function $f(p,T)$. However it is possible to make some general conclusion. As the value of p_1' approaches the value corresponding to merging of the flame front with the shock wave, that is as c_f tends to $c_{d\gamma}$ (point γ on Fig. 3), the velocity $c_{f\,chem}$ will necessarily remain lower than c_f. In the case when in a gasdynamically possible solution c_f behind the rarefaction wave tends to zero (point A on Fig. 3) the corresponding value of $c_{f\,chem}$ will remain finite, if the induction time as determined by formula (5.1) will not tend to infinity.

On the other hand, the slope of the curve $c_{f\,chem} = f(p_1')$ at the point corresponding to the initial state of the unburnt gas (point O on Fig. 3) may overtake any limit if the function $f(p,T)$ at this state has a very sharp dependence on its arguments.

It follows from the previous considerations that the dependence on $c_{f\,chem}$ on p_1' may be of the form illustrated by the curve $A_1\gamma_1$ on Fig. 3. This will be in case, for instance, when the function $f(p,T)$ is determined by the Arrhenius law with sufficiently high activation energy E.

Since the curve $A_1\gamma_1$ intersects the curve $A\gamma$ more than only at one point O, which corresponds to the initial state, it follows that in such cases the sudden splitting of the flame front is possible, and as a consequence, that the detonation wave with a two-front structure is unstable.

The sufficient criterium for this instability
is evidently the following condition

$$\left(\frac{dc_{f\,chem}}{dp_1'}\right)_{p_1'=p_{10}} > \left(\frac{dc_f}{dp_1'}\right)_{p_1'=p_{10}}$$

which has been formulated in $[31]$.

In following the analytical expressions for the
derivatives entering this condition will be given and thus an
analytical criterium sufficient for the instability of a detonation wave will be given.

Let us notify the substantially nonlinear mecha-
nism of the considered instability : the finite perturbations of
the flow – the variation of the flame front velocity and emission
by it of shock waves and centred rarefaction waves – may suddenly proceed as a result of arbitrarily small initial disturbances.

This instability may be considered also as a pa-
rametric one, depending on the values of the gas parameters a-
head of the detonation wave and on the parameters which enter in
the expression for induction time. Thus for instance if the parameters ahead of a detonation wave with a two-front structure
gradually vary (quasistatically) this detonation wave may suddenly lose its stability.

The selfsimilar flow which generates immediately
after the loss of stability will exist only in the region which

is not disturbed by the perturbations reflected from the head
shock of the detonation wave. The further development of the
flow field proceeds as a result of many complex reflections and
transmissions of emitted waves during their mutual interactions,
and interactions with shock wave, flame front and eventually
with boundaries of the flow. Also the possibility must be admit-
ted that in the course of the flow development the flame front
may again become instable and generate a new system of waves.

Therefore on the basis of above considera-
tions it would be hard to make definite conclusions concerning
the final behaviour of the detonation wave in the nonlinear theory ; the possible exception may be the following : if the induction time in the solution with emission of rarefaction wave
into unburnt gas is considerably larger than the characteristic
time for the flow corresponding to undisturbed solution, one can
say that the detonation practically stops.

Some general conclusions concerning the behaviour
of the perturbed detonation wave will be made later, however, on
ly in linear approximation or on the basis of numerical results.

In those cases in which apart from solution corre
sponding to steady two-front structure of detonation wave, there
exists one or more other solutions with splitting of initial dis
continuity on the flame front, the question arises about what
kind of perturbations will lead to a certain form of solution.
To answer this question in all generality is extremely difficult.

The paper [30] gives numerical examples of behaviour in time of stable and unstable normal detonation waves.
The calculations were made by finite difference method. For the
combustible mixture and for combustion products the following
simple equations of state were taken

$$p = \frac{1}{3} (\varrho - 1), \quad p = \frac{1}{3} \varrho^3$$

(p and ϱ are normalized by their values ahead of the shock).

The function f , which determines the induction
time was assumed to be constant and depending on the state of the
particle immediately behind the front shock according to formulae

$$f = \frac{1}{\tau} = 0 \qquad \varrho < \varrho_H = 1,59$$

$$f = \frac{1}{\tau} = k \left(\frac{\varrho}{\varrho_1} \right)^{\alpha} \qquad \varrho \geq \varrho_H$$

Here ϱ_1 is the density behind the front shock,
k - proportionality factor.

As the preliminary analysis has shown under the
above assumptions the combustion front in a normal detona-
tion wave is stable when $\alpha < 7,77$ and unstable - when $\alpha > 7,77$.

The calculations of flow were made with initial
data corresponding to a steady normal detonation wave with $\alpha = 0$
and $\alpha = 15$. Fig. 5a illustrates the density profile in this

steady wave obtained by the numerical procedure used. With $\alpha = 0$
this profile does not change as the wave propagates if the initial density distribution is perturbed by a small negative perturbation. With a small positive perturbation one may observe
small oscillations of density in the region between two fronts
(Fig. 5b).

　　　　　With $\alpha = 15$ a small negative perturbation of
density led to the emission ahead from the combustion front of
an intense rarefaction wave which weakend the front shock in
such a manner that the density behind it fell below the value
ϱ_μ and consequently the combustion stopped (Fig. 5c). In the
case of a positive initial perturbation a compression shock has
been developed ; this shock overtook the front shock increasing
its strength whereupon the strength of this shock began to decrease rapidly (Fig. 5d).

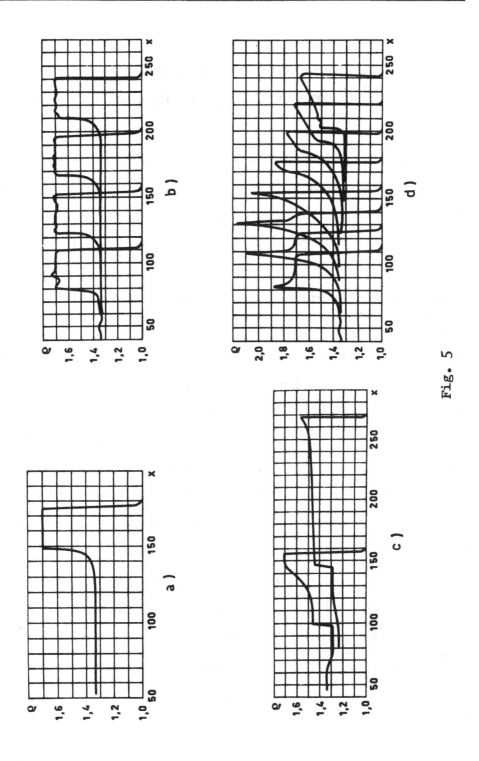

Fig. 5

3.2. LINEAR STABILITY THEORY OF DETONATION WAVE
WITH TWO–FRONT STRUCTURE

Let us consider again the one–dimensional gas flow with plane detonation wave having two–front structure (Fig. 1). The variations of velocity, pressure and density of the gas in each of the three regions, separated one of the other by adiabatic shock s and heat release front f are described by equations of adiabatic gas flow (Pt. 1, Eqs (2.1)). Let the subscript ∞ denote the values ahead of adiabatic shock, the subscript 1 – the values inside the layer between two fronts, and let the values behind the heat release front be without any subscripts.

The parameters of the gas on both sides of the adiabatic shock must be related by conditions (2.3, Pt. 1) with $Q = 0$ and with upper sign in front of the root; on both sides of the heat release front the same conditions must hold but with $Q \neq 0$ and with lower sign in front of the root.

Let the initial state (with parameters denoted by subscript 0) be the steady structure of the wave in the coordinate system in which the velocity of wave propagation equals zero. We shall consider nonsteady motions arising when this wave interacts with perturbations approaching it from the rear side [32].

For a slightly disturbed flow in the layer between two fronts the gas parameters inside this layer will be
presented in the form

$$v = v_{10} + \delta v_1 \qquad p = p_{10} + \delta p_1 \qquad \varrho = \varrho_{10} + \delta \varrho_1 .$$

To the linear approximation the gas dynamics
equations give

$$\delta v_1 = v_{10}(\mathcal{F} + \mathcal{G}) \qquad \delta p_1 = \varrho_{10} v_{10} a_{10}(-\mathcal{F} + \mathcal{G})$$

$$\delta \varrho_1 = \frac{\varrho_{10} v_{10}}{a_{10}}(-\mathcal{F} + \mathcal{G}) + \varrho_{10} \mathcal{H}$$

(5.2.1)

where the functions $\mathcal{F}$, $\mathcal{G}$, $\mathcal{H}$ depend correspondingly on one
characteristic variable

$$\xi = r - (v_{10} - a_{10})t , \quad \eta = r - (v_{10} + a_{10})t , \quad \zeta = r - v_{10}t .$$

The adiabatic shock conditions give after their
linearization the following relations between values of func-
tions $\mathcal{F}$, $\mathcal{G}$ and $\mathcal{H}$ on the front (in linear approximation at
$r = 0$) and the front velocity c_δ

$$\mathcal{G}_\delta = -\lambda \mathcal{F}_\delta \qquad \mathcal{H}_\delta = \sigma \mathcal{F}_\delta \qquad \frac{c_\delta}{v_{10}} = \varkappa \mathcal{F}_\delta . \qquad (2.2)$$

Here λ , σ , $\varkappa$ are functions of Mach number of oncoming
stream M_∞ and of adiabatic exponent γ .

The quantity $\lambda = -\dfrac{\mathcal{G}_\delta}{\mathcal{F}_\delta}$ gives the measure for the variation of pressure perturbation as it reflects from the shock and is usually called the reflection coefficient of small perturbations from the shock.

Let us emphasize that with given perturbations approaching the shock from the rear along the characteristics ξ = const (i.e. with given $\mathcal{F}$) the perturbations outgoing from the shock along the characteristics η = const and ζ = = const (i.e. functions $\mathcal{G}$ and $\mathcal{H}$) and the shock velocity c_{δ} are completely determined.

On the trajectory of a gas particle the coordinates of the shock r_{δ} and of the combustion front r_f are related by the condition

$$r_f(t+\tau) = r_{\delta}(t) + \int_t^{t+\tau} \upsilon dt$$

where τ is the induction time.

From this relation and from the determination of τ to linear approximation (assuming small variations of τ when changes of p and T are small) we obtain

$$r_f(t+\tau_0) = r_{\delta}(t) + \upsilon_{10}\tau_0 + \upsilon_{10}\int_t^{t+\tau_0}\left(\frac{\delta\upsilon_1}{\upsilon_{10}} - \frac{\partial\ln f}{\partial\ln p}\bigg|_0\frac{\delta p_1}{p_{10}} - \frac{\partial\ln f}{\partial\ln T}\bigg|_0\frac{\delta T_1}{T_{10}}\right)dt$$

(2.3)

Using the expressions for gas parameters (2.1) and shock conditions (2.2) the relation (2.3) after differentiation in respect to t may be transformed in such a manner that it will include only the values of gas parameters ahead of the combustion front and the velocity c_f of this front :

$$\frac{c_f}{v_{10}} = (1-\mu M_{10}^2)\frac{\delta v_{1f}}{v_{10}} + (1-\mu)\frac{\delta p_{1f}}{\varrho_{10}a_{10}^2} - \left[\frac{1}{2} + \frac{\mu M_{10}^2 + \frac{\gamma-1}{2}M_\infty^2}{2(M_\infty^2 M_{10}^2 - 1)}\right]\frac{\delta s_{1f}}{c_p} - n\tau_0\frac{\partial}{\partial t}\frac{\delta s_{1f}}{c_p}$$

$$(2.4)$$

Here $\mu = (\gamma-1)\left.\frac{\partial \ell n f}{\partial \ell n T}\right|_0 + \gamma\left.\frac{\partial \ell n f}{\partial \ell n p}\right|_0$, $n = \left.\frac{\partial \ell n f}{\partial \ell n T}\right|_0$. The relation
obtained is just the necessary additional condition to the con-
servation laws on the heat release front.

We turn now to the conditions following
from the conservation laws. Let us introduce parameter Δ which
characterizes the deviation of the unperturbed detonation wave
from the Chapman-Jouget regime by the formula (the parameter Δ
is simply connected to parameter ε used in Pt. I):

$$\left(\frac{a_{10}^2}{v_{10}} - v_{10}\right)^2 - 2(\gamma-1)Q = v_3^2\Delta^2$$

For any value of Δ and slightly perturbed flow
the conditions on the heat release front will have the following
form : at $r = v_{10}\tau_0$

$$\delta v = v - v_0 = \frac{1}{\gamma+1}\left\{\left(1+\frac{1}{M_{10}^2}\right)c_f + \left(\gamma-\frac{1}{M_{10}^2}-\frac{\gamma-1}{M_{10}}\right)v_{10}\mathcal{F} + \right.$$

$$+\left(\gamma-\frac{1}{M_{10}^2}+\frac{\gamma-1}{M_{10}}\right)v_{10}\mathcal{G} - \frac{1}{M_{10}^2}v_{10}\mathcal{H} + v_3\Delta -$$

$$-\sqrt{v_3^2\Delta^2 + 2v_{10}\left(\frac{1}{M_{10}^2} - 1\right)\left[\left(1 + \frac{1}{M_{10}^2}\right)c_f - \left(1 + \frac{1}{M_{10}^2} + \frac{\gamma-1}{M_{10}^2}\right)v_{10}\mathfrak{F} -}$$

$$\overline{\left. - \left(1 + \frac{1}{M_{10}^2} - \frac{\gamma-1}{M_{10}^2}\right)v_{10}\mathfrak{G} - \frac{1}{M_{10}^2}v_{10}\mathcal{H}\right]\right\}}$$

(2.5)

$$\frac{\delta p}{\varrho_0 v_0} = -\delta v - \left(1 - \frac{v_0}{v_{10}}\right)c_f + \left[2 - \frac{1}{M_{10}} - M_{10} - (1 - M_{10})\frac{v_0}{v_{10}}\right]v_{10}\mathfrak{F} +$$

$$+ \left[2 + \frac{1}{M_{10}} + M_{10} - (1 + M_{10})\frac{v_0}{v_{10}}\right]v_{10}\mathfrak{G} + v_{10}\left(1 - \frac{v_0}{v_{10}}\right)\mathcal{H}$$

$$\frac{\delta p}{\varrho_0} = -\frac{\delta v}{v_0} + \left(\frac{v_{10}}{v_0} - 1\right)\frac{c_f}{v_{10}} + (1 - M_{10})\mathfrak{F} + (1 + M_{10})\mathfrak{G} + \mathcal{H}.$$

Let us consider at first the case when Δ is not small. Then in expressions (2.5) the quantities $\frac{c_f}{v_{10}}$, $\mathfrak{F}$, $\mathfrak{G}$, $\mathcal{H}$ may be assumed to be small compared with Δ^2 , and the flow behind the heat release shock may be treated in linear approximation, i.e. one may put

$$\delta v = v_0(F + G), \quad \delta p = \varrho_0 v_0 a_0(-F + G), \quad \delta\varrho = \frac{\varrho_0 v_0}{a_0}(-F + G) + \varrho_0 H.$$

After linearization of the square root in the first of Eqs. (2.5) these three equations and the equation (2.4) give four linear conditions relating the velocity of the heat release front c_f and three outgoing from the front

perturbations $\mathfrak{I}$, G , H to the three coming to the front perturbations $\mathcal{G}$, $\mathcal{H}$, F .

In the absence of any coming perturbations $\delta\delta_{1f} = 0$
$\delta p_{1f} = -\varrho_{10} a_{10} \delta v_{1f}$ and $\delta p_f = \varrho_0 a_0 \delta v_f$. The gasdynamic conservation
laws give in this case

$$C_f = B\left(M_\infty, \gamma, \Delta\right) \delta v_{1f} .$$

On the other hand under the same conditions from
the "kinetic" relation (2.4) follows

$$C_{f\,chem} = \left(1 - M_{10}\right)\left(1 + \mu M_{10}\right)\delta v_{1f} .$$

Thus, in accordance with the results of the preceding section the condition for instability of the heat release
front is found in analytical form :

$$\frac{C_{f\,chem}}{\delta p_1} > \frac{C_f}{\delta p_1} \qquad \left(1 - M_{10}\right)\left(1 + \mu M_{10}\right) < B\left(M_\infty, \gamma, \Delta\right) .$$

This condition which fulfilled the combustion
front may get new velocity emitting shock wave or centred rarefaction wave.

On Fig. 6 the solid lines represent the stabil-
ity boundaries as dependence of $\dfrac{E}{R T_\infty}$ on M_∞ for different values of $\dfrac{Q}{c_p T_\infty}$. Left from the curves the flow is unstable due to
possible instantaneous change of the combustion front velocity.
The bottom limiting curve corresponds to Chapman-Jouget regimes:

on this curve $Q = Q_J$ or $\Delta = 0$.

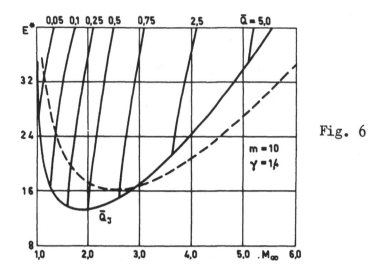

Fig. 6

In linear approximation with no perturbations
coming to the heat release front nonzero solutions exist only
on the stability boundary as determined before. It is obvious
that the reflection and transmission coefficients for the coming
to the front perturbations become infinite on the stability
boundary. In all other cases the instability may arise in the
usual way as a result of gradual amplification of initial perturbation.

       To investigate the stability in linear approxi-
mation let us analyze the system of conditions on the heat release front : condition (2.4) and linearized conditions (2.5).
We assume the detonation wave to interact with perturbations
approaching it from the rear. Then the perturbations $\mathcal{G}$ and $\mathcal{H}$
coming to the heat release front from the fore side are formed

as result of reflection from the adiabatic shock of perturbation $\mathfrak{F}$ propagating from the heat release front ahead of it that is

$$\mathcal{G}_{f,t} = -\lambda \mathfrak{F}_{f,t-\delta_1} \,, \quad \mathcal{H}_{f,t} = \sigma \mathfrak{F}_{f,t-\delta_2} \,, \quad \mathcal{H}'_{f,t} = -\sigma \left(\frac{a_{10}}{v_{10}} -1 \right) \mathfrak{F}'_{f,t-\delta_2}$$

where $\delta_1 = \dfrac{2 M_{10}}{1-M_{10}^2} \tau_0 \,, \quad \delta_2 = \dfrac{\tau_0}{1-M_{10}} \,.$

Using this relations we get the following equation determining the function $\mathfrak{F}$

$$\mathfrak{F}_{f,t} = \lambda' \mathfrak{F}_{f,t-\delta_1} + \lambda'' \mathfrak{F}_{f,t-\delta_2} + \lambda''' v_{10} \tau_0 \mathfrak{F}'_{f,t-\delta_2} + K \Delta F_f \quad (2.6)$$

where the quantities λ', λ'', λ''', K are fractionally linear functions of m and n with coefficients depending in the known way on parameters of the oncoming stream.

As expected all these quantities rise unlimitely when the parameters of the heat release front approach their stability boundary. In the limiting case $\Delta \ll 1$ this boundary is determined by the formula

$$\mu = \frac{\gamma M_{10}^2}{(1+M_{10}^2)(1-M_{10})} \,.$$

On Fig. 6 to this formula corresponds the curve $\bar{Q} = \bar{Q}_3$ or $\Delta = 0$. Let us remind that the Eq. (2.6) is valid only if $\mathfrak{F} \ll \Delta^2$, consequently the coming perturbation must satisfy the condition $F \ll \Delta$.

3.3. DEVELOPMENT OF OSCILLATIONS ASSOCIATED WITH THE ATTENUATION OF A DETONATION WAVE

With small values of Δ it is not possible to linearize the conservation equations on the heat release front and the equations of combustion products flow. The reason for this is that for small values of Δ it is important to consider the incident perturbations of order Δ and the velocity of the gas behind the front differs from the velocity of sound also by an amount of order Δ . In particular, to analyze the approach of a slightly overdriven detonation wave to the Chapman–Jouget regime we must assume the perturbations attenuating the detonation wave to be of the order Δ .

Let us consider Eqs. (2.5) in the case when the perturbation approaching the heat release front is of order Δ . The first condition (2.5) shows that in this case the values $\dfrac{c_f}{v_{10}}$ and $\mathcal{F}$ (and consequently in the problem under consideration – the functions $\mathcal{G}$ and $\mathcal{H}$ as well) must be of the order Δ . Neglecting the terms of higher order the first condition (2.5) gives

(3.1a)
$$ v - v_3 = - \sqrt{ v_3^2 \Delta^2 + 2 v_{10}^2 \left(\frac{1}{M_{10}^2} - 1 \right) \left[\left(1 + \frac{1}{M_{10}^2} \right) c_f - \right. } $$

$$-\left(1+\frac{1}{M_{10}^2}+\frac{\gamma-1}{M_{10}}\right)\mathfrak{F}-\left(1+\frac{1}{M_{10}^2}-\frac{\gamma-1}{M_{10}}\right)\mathfrak{G}-\frac{1}{M_{10}^2}\mathcal{H}\bigg].$$

$$(3.1)$$

In accordance with results of Sec. 13 we conclude that the perturbations behind the heat release front, which are of order Δ , represent a simple wave approaching the heat release front ; the perturbations leaving the front in downstream direction are of order Δ^2.

Thus the left-hand side term in condition (3.1) may be considered as given. If the perturbations incident on the heat release front from behind are of order Δ this condition replaces the condition obtained by linearization of the first Eq. (2.5) which is valid in the case when incident perturbations are of higher order.

Evidently both these conditions differ only by terms which characterize the disturbances incident on the heat release front from behind. Therefore in the case under consideration as in the previous one the equation (2.6) may be used with the same coefficients λ' , λ'' , λ''' but putting in its last term instead of $\Delta F_{f,t}$ the quantity

$$\frac{1}{2(\gamma+1)}\left[\Delta^2-(\gamma+1)^2\frac{(\upsilon_3-\upsilon)^2}{\upsilon_3^2}\right].\qquad(3.2)$$

Hence, let us consider a simple wave incident

on the heat release front from behind. The formulae describing

this wave

$$r = (\upsilon - a)t + \Phi(\upsilon), \qquad \upsilon + \frac{2a}{\gamma - 1} = const$$

to the approximation considered may be replaced by the follow-

ing

$$r = \frac{\gamma + 1}{2}(\upsilon - \upsilon_3)t + \Phi(\upsilon), \qquad \upsilon + \frac{2a}{\gamma - 1} = \upsilon_3 + \frac{2a_3}{\gamma - 1}$$

If the function $\Phi(\upsilon)$ has a finite derivative

at $\upsilon = \upsilon_3$, than within the same accuracy

(3.3) $$r - r_0 = \frac{\gamma + 1}{2}(\upsilon - \upsilon_3)(t - t_0)$$

where $r_0 = \Phi(\upsilon_3)$, $t_0 = -\frac{2}{\gamma + 1}\Phi'(\upsilon_y)$, i.e. the wave can be

regarded as centred.

Taking the value of $\upsilon - \upsilon_3$ at $r = \upsilon_{10}\tau_0$ from

(3.3) the expression (3.2) may be given the form

(3.4) $$\frac{\Delta^2}{2(\gamma + 1)}\left[1 - \left(\frac{t_0}{t - t_0}\right)^2\right]$$

After substituting (3.4) instead of $\Delta F_{f,t}$ in

the Eq. (2.6) it is easy to determine the function $\mathfrak{F}$ and con-

sequently the other functions in the layer between two fronts

inside of each region separated by characteristics belonging to

different families. The flow perturbations behind the heat release front that are superimposed on the incident simple wave
and which are of order Δ^2 may be found, in the event it is necessary, from conditions (2.5).

Let us now analyze the asymptotic behaviour of
the solution to the equation (2.6) at $t \longrightarrow \infty$. We assume for
$\mathfrak{F}_{f,t}$ the form $\mathfrak{F}_{f,t} = \Delta^2 \left(c + c_1 e^{z\frac{t}{t_0}} \right)$, where $z = \beta + i\Omega$ and β
and Ω are real numbers. Substituting this expression into
Eq. (2.6) we find

$$c = \frac{K}{2(\gamma+1)(1-\lambda'-\lambda''')}$$

and after some rearrangements

$$\frac{(c_f)_\infty}{\upsilon_\infty} = \frac{(1+\gamma M_\infty^2)^2}{2(\gamma+1)^2(M_\infty^2+1)(M_\infty^2-1)} \Delta^2$$

The velocity of the detonation front according
to this formulae corresponds to the Chapman-Jouget regime.

In the solution of the homogeneous equation for
$\mathfrak{F}$ the complex constant z must satisfy the condition

$$\lambda' e^{-z\frac{\delta_1}{t_0}} + \left(\lambda'' + \frac{M_{10}}{1-M_{10}} \lambda''' z \right) e^{-z\frac{\delta_2}{t_0}} - 1 = 0$$

The solution to be stable all roots of this equation must lie to the left from the imaginary axis on the plane

of complex variable z .

However, it may be shown that this equation has an infinite set of roots with increasing Ω ; with the real parts of the roots become positive for sufficiently large Ω . This behaviour is easy to predict from Eq. (2.6) which contains along with the function $\mathcal{F}$ its derivative $\mathcal{F}'$ as well (the coefficient at $\mathcal{F}'$ being proportional to λ''').

The dependence on M_∞ (for $\gamma = 1.4$) of values $\dfrac{E}{RT_\infty}$ with smallest Ω is shown as dotted line on Fig. 6.

Let us emphasize that intense oscillations can originate during the initial stage of interaction of the detonation wave having a two—front structure with a rarefaction wave approaching it from behind. Figure 7 shows an example of the changes in the velocity of the front of a detonation wave as a consequence of its interaction with rarefaction waves of different intensity; the intensity of the wave is defined by the parameter t_0 – the time coordinate of the centre of the rarefaction wave as a fraction of the quantity τ_0 .

In a series of papers the stability of a plane detonation wave with extended heat release zone is considered respective to small transverse perturbations of different wave length and as a particular case – respective to longitudinal perturbations (see [33] , also [34] , where the results in this field obtained in USA are reviewed).

Qualitatively the results of these papers concern ing the longitudinal perturbations are similar to those describ- ed above for two-front models of the detonation wave. Of course the analysis of the model with extended heat release zone turns out to be very complicated and can be realized only with the use of numerical methods.

Numerical methods have been used also to get so- lutions of a number of nonlinear problems concerning the attenua tion of detonation wave by a rarefaction wave approaching it from behind.

In the paper [35] a problem is considered which we may treat as follows. A plane piston starts to move in a com bustible gas initially at rest in such manner that the develop- ing shock wave propagates with constant velocity exceeding the Chapman-Jouget velocity. To satisfy this requirement in the case of flow with heat release behind the shock the piston velocity must decrease gradually. When the reaction in the gas particle at the piston has been substantially completed the piston moves with constant velocity during a certain time interval, then sud denly the velocity changes to a value which is lower than that corresponding to Ch.-J.regime. Thus the stationary (in the coor dinate system related to the shock) overdriven detonation wave is overtaken by a centred rarefaction wave approaching it from behind.

These examples calculated by the method of char-

acteristics enable the authors of paper 35 to make conclusion that
with strong dependence of chemical reaction rate on the temperature, the intensity of the shock wave falls below the Ch.-J. value.

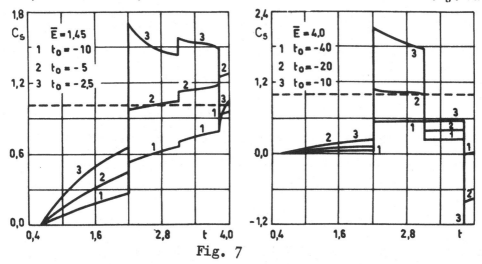

Fig. 7

In the case in which the reaction rate was considered as independent on the temperature the results of calculations did not
permit to make any definitive conclusion : during the time interval considered the intensity of the shock was continuously
decreasing but it still exceeded that of the Ch.-J. wave. Let
us note that in all investigated examples the distance of
the centre of the rarefaction wave from the rear front of the
detonation wave was of the order of few thicknesses of the chem
ical reaction zone, and the time from the beginning of interaction between the rarefaction wave and the detonation wave and
up to the end of calculations was only few times larger than
the time of the chemical reaction.

The paper [36] treats the same piston problems

as the paper [35] , the difference from the latter is that in
paper [36] after the piston velocity becomes constant corresponding to supercompressed detonation wave it gradually decreases to a new constant value which again corresponds to a
supercompressed wave. The calculations were made for a larger
time interval exceeding to several tens the chemical reaction
time. The results obtained show that when the piston velocity
changes from a larger value to a smaller one, than within certain range of defining parameters after some transition process,
a flow develops with shock wave that intensity is smaller, and
the structure of the detonation wave transforms into a new
steady state corresponding to the new piston velocity. However,
this does not happen in all cases. With other values of defining parameters the decrease of the piston velocity leads to the
development of periodic oscillatory propagation regimes of the
shock wave and of the following combustion zone. These oscillations may be very intense ; in the calculated examples the variation of the pressure behind the shock wave was larger than
one half of its value corresponding to the steady wave propagation. The transition to a new steady state occured in these
cases when the steady structure of the wave was stable to linear
approximation with respect to longitudinal perturbations ; oscillatory regimes developed in cases when the new structure was
unstable to linear approximation.

The paper [37] treat numerically the problem of

interaction between a detonation wave with an extended chemical
reaction zone described by the following kinetic equation

$$\frac{d\beta}{dt} = -K \sqrt{\beta} \exp\left(-\frac{E}{RT}\right)$$

(notations are the same as in Sec. 1.5) and the centred rarefaction flow approaching to the wave from rear.

Calculations were carried in Lagrangian variables
by the method of characteristics, as thermodynamic variables to
be determined the quantities $\frac{p}{\varrho^\gamma}$ and p were used.

Fig. 8 and Fig. 9 present calculated dependencies
on time of the front velocity of the shock wave followed by a
combustion zone. The graphs on Fig. 8 correspond to different
intensities of the rarefaction wave (parameter t_0 characterizes
the propagation time of the disturbance from the rarefaction wave
centre up to the reaction zone) and to that value of the activation energy E for which the oscillations developing in the flow
attenuate and the detonation wave tends to Chapman-Jouget regime.
The graphs on Fig. 9 correspond to the same value of the parameter t_0 but to different values of activation energy. With E
sufficiently high the amplitude with which the forder front of
detonation wave oscillates is increasing with time. The calculations made for the values of parameters close to the stability limit showed that the amplitude of oscillations stabilizes
with the time and regime of selfsubstaine oscillations of the

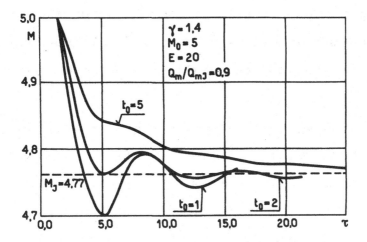

Fig. 8

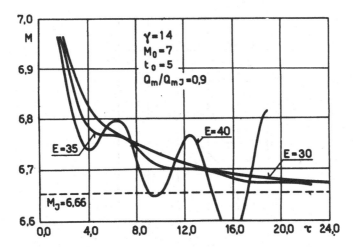

Fig. 9

detonation wave develops. With large values of E the velocity
of the forder front of the detonation wave during the initial
stage of interaction becomes so low that it may practically mean
inhibition of combustion behind the shock – the detonation de-
cays.

The solid line on Fig. 10 shows the stability
limit resulting from calculations (for $t_0 = 5$, nevertheless
the variation of t_0 to larger values does not influence this
limit). Three other lines correspond to analytical results
(linear theory of Sec. 3.2, linear theory according to criterium on page 119, according to paper $[38]$, where no account
was taken for flow variation between two fronts).

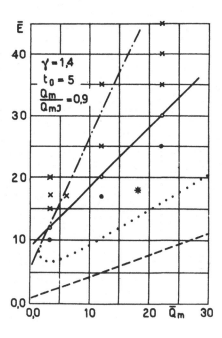

Fig. 10

3.4. POINT EXPLOSION IN COMBUSTIBLE GAS MIXTURE [39]

Let us assume that in an unbounded quiescent gas, in which exothermic chemical reactions are possible, an energy amount E_0 is released instantaneously along a plane, a straight line or at a point. Then a strong shock wave will start to move through the gas switching on chemical reactions with heat output. If one assumes the thickness of the combustion zone vanishingly small, as compared with shock wave radius $r_{_{\! j}}$, and the combustion to be complete, then the energy released by the combustion of gas mixture is

$$E_{chem} = \sigma_{_{\! v}} \varrho_1 r_{_{\! j}}^{\nu} Q \qquad \sigma_{_{\! v}} = \begin{array}{ll} 2 & \nu = 1 \\ \pi & \nu = 2 \\ \dfrac{4}{3}\pi & \nu = 3 \end{array}$$

During the initial period of motion when

$$r_{_{\! j}}^{\nu} \ll \frac{E_0}{\sigma_{_{\! v}} \varrho_1 Q}$$

the following condition $E_{chem} \ll E_0$ is satisfied, what means that the chemical reactions do not have any considerable influence on the gas motion. At later stages the chemical reactions may change substantionally the flow pattern.

For the infinitely thin detonation wave some results of the solution of this problem were considered in

Below we consider two other models of the gas
flow.

Let us first use the model 1a with induction time
τ defined by the formula

$$\tau = k\,p_{\,3}^{-n}\,\varrho_{\,3}^{-\ell}\,\exp\frac{E}{RT_{\,3}}\,.$$

Leaving aside the details of the flow behind the
front of shock wave let us consider roughly the qualitative variation of τ behind the shock, assuming $n > 0$; $\ell > 0$. With
large $p_{\,3}$ and $T_{\,3}$ the induction time is small. As the shock wave
attenuates and $p_{\,3}$ and $T_{\,3}$ in the gas particle decrease the time τ
increases and consequently the distance between the shock wave
and combustion front increases too – the combustion front sep
arates from the shock wave. Thus beginning from a certain moment the shock wave and the combustion zone cannot be considered as a single discontinuity surface – detonation wave.

Let us describe some results of analysis of detonation wave splitting during the initial stage of point explosion. Fig. 11a illustrates the gas flow pattern. In the region 1
gas is at rest, in the region 2 behind the shock the gas moves
adiabatically. The gas burns crossing the combustion wave, thus
the region 3 is filled by combustion products. The splitting of
detonation wave is sketched qualitatively on Fig. 11b.

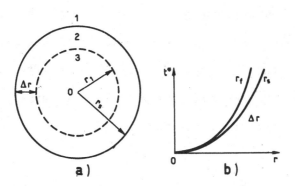

Fig. 11

The calculations showed that the time t when the shock and the combustion front begin to diverge quickly increases with increasing E_0 and decreases with increasing E. These effects are illustrated on Fig. 12 in the case of spherical symmetry.

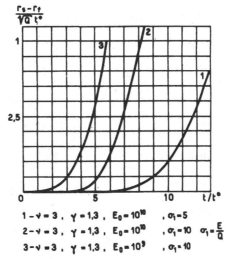

1 - $v = 3$, $\gamma = 1,3$, $E_0 = 10^{10}$, $\sigma_1 = 5$
2 - $v = 3$, $\gamma = 1,3$, $E_0 = 10^{10}$, $\sigma_1 = 10$ $\sigma_1 = \frac{E}{a}$
3 - $v = 3$, $\gamma = 1,3$, $E_0 = 10^{9}$, $\sigma_1 = 10$

Fig. 12

Now we turn to some features of the solution for
the model in which the chemical reaction with heat release in
the flow behind the shock wave is switched on after the induction time (model 3). We take the equations governing the chemical reactions in the Arrhenius form

$$\frac{dc}{dt} = k\, p^n \varrho^\ell \exp\left(-\frac{E\varrho}{p}\right)$$

$$\frac{d\beta}{dt} = -k_1 \beta^{m_1} p^{n_1} \varrho^{\ell_1} \exp\left(-\frac{E_1\varrho}{p}\right) + k_2 (1-\beta)^{m_2} p^{n_2} \varrho^{\ell_2} \exp\left(-\frac{E_2\varrho}{p}\right)$$

For c the condition $c = 0$ on the shock wave
holds. If $c = 1$ it means that the induction time is over and
the exothermic chemical reaction starts.

The calculations showed that due to large negative gradients of temperature, pressure and density of the gas
in the expansion region behind the shock wave the combustion
zone separates from the shock wave. The induction time increases
considerably as the shock propagates in spite of the fact that
the shock still remains sufficiently strong $(p_3 > 50\, p_1)$.

As a consequence the detonation wave splits into
an adiabatic shock wave and flame front.

The calculations showed also that the concentra-
tion β is strongly influenced by the inverse reaction which is
oftenly neglected when modelling description of the gas flow be-

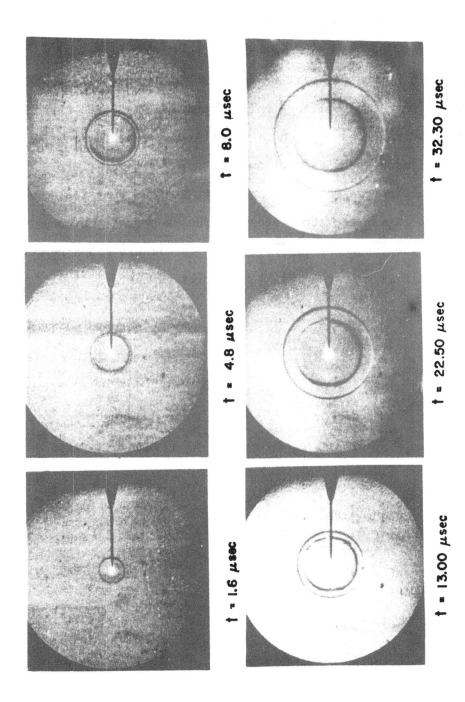

t = 1.6 μsec   t = 4.8 μsec   t = 8.0 μsec

t = 13.00 μsec   t = 22.50 μsec   t = 32.30 μsec

Fig. 13

hind the detonation wave is used.

Due to high temperature in the vicinity of the explosion centre the gas mixture does not burn completely and the thermal energy is released only partially. The ratio of the heat released under assumption of model 3 to the heat which would be released under assumption of model 2 with complete burning equals approximately to one-half. At that time the combustion zone is already separated from the shock and the amount of the energy released by the combustion in all disturbed region is still small as compared with the explosion energy.

REFERENCES

[1] ЛАНДАУ Л.Д, ЛИФШИЦ И.М.–Механика сплошных сред. М.1962

[2] СЕДОВ Л.И.–Механика сплошной среды.т.I, Изд. "Наука".М.1970.

[3] ЧЕРНЫЙ Г.Г. Асимптотическое поведене плоской детонационной волны.Докл.АН СССР, № 3, 1967.

[4] ЛЕВИН В.А, ЧЕРНЫЙ Г.Г. Асимптотические законы поведения детонационных волн.Прикладн.матем. и механика, вып.6, 1967.

[5] KOROBEINIKOV V.P.– The problem of point explosion in a detonating gas. Astronautica Acta 14, № 5, 1969.

[6] ГИЛИНСКИЙ С.М., ТЕЛЕНИН Г.Ф., ТИНЯКОВ Г.П.– Метод расчета сверхзвукового обтекания затупленных тел с отошедшей ударной волной.Изв. АН СССР, ОТН, Механика и машиностроенте, 1964, № 4.

[7] ГИЛИНСКИЙ С.М., ЗАПРЯНОВ Э.Д., ЧЕРНЫЙ Г.Г.– Сверх звуковле обтекание сферы горючей смесью газов. Изв.АН СССР,Механика жидкости и газа,1966,№ 5.

[8] ГИЛИНСКИЙ С.М., ЧЕРНЫЙ Г.Г.– Сверхзвуковое обтекание сферы с учетом времени задержки воспламенения.Изв.АН СССР.Механика жидкости и газа, 1968, № 1.

[9] ГИЛИНСКИЙ С.М.– Расчет течения горючей смеси перед затупленным телом.Изв.АН СССР, Механика жидкости и газа, 1968, № 2.

[10] ГИЛИНСКИЙ С.М.– Расчет горения водорода в воздухе за отошедшей ударной волной при сверхзвуковом движении сферы.Изв.АН СССР, Механика

жидкости и газа, № 4,1969.

[11] СТУЛОВ В.П., ТУРЧАК Л.И.- Сверхзвуковое обтекание
сферы гремучей смесью.Изв.АН СССР, Механика
жидкости и газа, 1968, № 6.

[12] ГИЛИНСКИЙ С.М., ШКАДОВА В.П.- Численное решение
некоторых двумерных задач внешней аэродина-
мики при наличии горения.Аннотация доклада
II Междунаодн.коллокв.по газодинамике взрыва
и реагирующих систем.г.Новосибирск, 1969.

[13] ШКАДОВА В.П.- Околоравновесное обтекание тел вра-
щения сверхзвуковым потоком воздуха.Изв.АН
СССР, Механика жидкости и газа, 1969, № 1.

[14] SHERMAN M.A. - Radiation coupled chemical nonequilibrium
normal shock waves. J.Quant.Spectrosc.Radiat. Transfer.
v.8, 1968.

[15] GALLOWAY J., SICHEL M. - Hypersonic blunt body flow of
H_2-O_2 mixtures. Astronautica Acta v.15, pp.89 - 105.

[16] CONTI R.J. - A theoretical study on nonequilibrium blunt-
body flows. J.Fluid Mech. 24, 1, 1965.

[17] MORETTI G. - A new technique for the numerical analysis of
nonequilibrium flows. AJAA Journal, N.2, 1965.

[18] DEGROAT J.J., ABBETT M.J. - A computation of combustion
of methane. AJAA Journal, N.3, 1965.

[19] ГИЛИНСКИЙ С.М., ЗАПРЯНОВ Э.Д.- О переходие сверхзвукового течения горючей смеси газов к режиму Чепмена-Жуге, Изв.АН СССР, Механика жидкости и газа, 1967, № 3.

[20] "Движение тел с большой скоростью в химически ак-
тивных газах".Отчет Института механики МГУ,
№ 797, 1967 г.

[21] ГИЛИНСКИЙ С.М., МЕДВЕДЕВ С.А.- Движение тел с
большой скоростью в горючих смесях газов.

Аннотации докладов III Всесоюзного съезда по теоретич.и прикладн.механике, 1968.

[22] ЧУШКИН П.И.- Сверхзвуковое обтекание тел горючим газом.Физ.горения и взрыва, № 2, 1969 г.

[23] ЧУШКИН П.И.- Горение в сверхзвуковых потоках, обтекающих различные тела.Ж.Вычисл.матем.и матем.физ.1968,8, № 5.

[24] ЧЕРНЫЙ Г.Г.- Сверхзвуковое обтекание тел с образованием фронтов детонации и горения. В кн. "Проблемы гидродинамики и механики спошной среды", М , 1969 г.

[25] A technique for studying supersonic combustion in the vicinity of a hypersonic missile. Nat.Bur.Stand.Techn. News Bull. V.44, N.11, 1960.

[26] BEHRENS H., STRUTH W., WECKEN F. - Studies of hypervelocity firings into mixtures of hydrogen with air or with oxygen. Tenth Symposium (Inter.) on Combustion, 1965.

[27] GILINSKII S.M., CHERNYI G.G. - High velocity motion of solids in combustible gas mixtures. Second international colloquim on gasdynamics of explosions and reactive systems, Novosibirsk, 1965.

[28] БАМ-ЗЕЛИАОВИЧ Г.М.- Рачпад произвольново разрыва в горючей смеси.Сборн.статей № 4 "Теоретич. гидромехн". М.1949.

[20] СЕДОВ Л.И.- Методы подобия и размерности в механике.Изд."Наука", М.1967.

[30] ИЛЬКАЕВА Л.В., ПОПОВ Н.В.- Гидродинамические решения для возмущенний неустойчивой детонационной волны. Физика горения и взрыва, № 3,1965.

[31] ЗАЙДЕЛЬ Р.М., ЗЕЛЬДОВИЧ Я.Б.- Одномерная неустойчивость и затухание детонации.Журнал Прикладн. механ.и технич.физики.№ 6, 1963.

[32] ЧЕРНЫЙ Г.Г.- Возникновение колебаний при волн де-

тонации.Прикладн.матем.и механика.23б вып.3, 1969.

[33] ПУХНАЧЕВ В.В.- О устойчивости детонации Чепмена-Жуге.Журнал Прикладн.механ.и технич.физики. № 6, 1963.

[34] STREHLOW R.A. - Gas phase detonations: recent developments. Combustion and Flame. Vol.12, N.2, 1968.

[35] STREHLOW R.A., HARTUNG W.E. - On the early relaxation of an overdriven detonation wave. Combustion and Flame. Vol.9, N.4, 1965.

[36] FICKET W., WOOD W.W. - Flow calculation for pulsating one-dimensional detonation wave. Combustion and Flame. Vol.9, N.4, 1965.

[37] МЕДВЕДЕВ С.А.- Об ослаблении пересжатых детона-ционных волн с конечной скоростью реакции. Известия АН СССР, Механика жидкости и газа, № 3, 1969.

[38] ЛЕВИН В.А.- О переходе плоской пересжатой детона-ционной волны к режиму Чепмена-Жуге.Известия АН СССР.Механика жидкости и газа.№ 2, 1968.

[39] КОРОБЕЙНИКОВ В.П., ЛЕВИН В.А.- Сильный взрыв в горючей смеси газов, Известия АН СССР.Механи-ка жидкости и газа, № 6, 1969.

CONTENTS

Printed in the United States
By Bookmasters